Praise for

The Biology of W

The Biology of Wonder is a wonderfully eclectic and wide-ranging book that clearly shows that all beings and landscapes on our fascinating and magnificent planet are deeply interconnected. In the spirit of personal rewilding, Professor Weber writes about interbeing, ecological commons, first-person ecology, and non-duality in ways that will make sense to readers with different interests, and his ideas about "poetic ecology" show clearly that we are not alone — indeed, we are one of the gang — and must not behave as if we are the only show in town.

—MARC BEKOFF, University of Colorado, and author,
Rewilding Our Hearts: Building Pathways of Compassion and Coexistence

Weber moves biology beyond reductionism into a new expanded view of life that includes not only reductionism itself, but also the interactive cooperation, beauty, and vital force that complete the picture of our living world.
—DAVID EHRENFELD, Distinguished Professor of Biology at Rutgers, and author, *The Arrogance of Humanism* and *Becoming Good Ancestors: How We Balance Nature, Community, and Technology*

The Biology of Wonder is a wonder. Schrodinger asked "What is Life" with brilliance, but misses "What IS life". Weber sees aliveness as functional wholes, self-creative, and self-generating, that co-create their worlds. The "aliveness" of all life, emotional, sentient, adgential, interested, co-mingled, "entangled with all of life", reorients us scientifically, poetically and morally, from the rich but insufficient reductionism Schrodinger helped spearhead.
—STUART KAUFFMAN FRSC, Emeritus Professor, University of Pennsylvania

Written with poetic elegance and interwoven with a rich vein of personal narrative, this extraordinary book takes the central idea of the subjectivity and interior life of all living beings and gives it concreteness by grounding it in the findings of modern biology. In articulating the Laws of Desire inherent in all organic life, it goes far toward reframing the debate about the relationship between mind and body.
—SHIERRY WEBER NICHOLSEN, author,
The Love of Nature and the End of the World

In Andreas Weber's vision, nature is beautiful, and ecology is poetry. Follow his beautiful words into a science that investigates the Earth as a breathing, sensitive planet that welcomes us with story and song.

—DAVID ROTHENBERG, author,
Survival of the Beautiful and *Bug Music*

The Biology of Wonder guides us toward discerning that value, meaning, experience, creativity, and freedom exist within and constitute the living world. Previously dismissed as "romantic," this viewpoint, at once clearheaded and compassionate, is tenaciously represented by Andreas Weber as deep realism. To come to grips with the understanding he communicates — to recognize the ubiquity of subjectivity in the world and the feeling-unity of the human with all creation — is to glimpse what biodiversity destruction heralds for the human soul. The work of protecting and restoring nature simultaneously recovers and rescues our innermost being.

—EILEEN CRIST, coeditor,
Keeping the Wild: Against the Domestication of Earth.

The Biology of Wonder is a wonderful biology, even a transformational one. Prof. Weber leads us into a radiant world which is sensuous, interconnected and always communicating in a bio-poetical symphony. Had we ears to hear the language and eyes to see the vision revealed in this book we would surely be made more alive and deeply thankful. This is more than a book; it is a revelation, and it joins the very few works I would take into the wilderness with me. Beautiful, wise, and grounded, I am grateful as much for the vision Prof Weber elucidates as for the love with which he clearly expresses it all.

—KALEEG HAINSWORTH, author,
An Altar in the Wilderness

Grounded in science, yet eloquently narrated, this is a groundbreaking book. Weber's visionary work provides new insight into human/nature interconnectedness and the dire consequences we face by remaining disconnected.

—RICHARD LOUV, author,
The Nature Principle and *Last Child in the Woods*

THE
BIOLOGY
OF WONDER

ALIVENESS, FEELING, AND THE
METAMORPHOSIS OF SCIENCE

ANDREAS WEBER

new society
PUBLISHERS

Cover design by Diane McIntosh. Watercolor image ©istock -stereotype

Printed in Canada. First printing January 2016.

Paperback ISBN: 978-0-86571-799-2 eBook ISBN: 978-1-55092-594-4

Funded by the Government of Canada | Financé par le gouvernement du Canada | Canada

Inquiries regarding requests to reprint all or part of *The Biology of Wonder* should be addressed to New Society Publishers at the address below. To order directly from the publishers, please call toll-free (North America) 1-800-567-6772, or order online at www. newsociety.com

Any other inquiries can be directed by mail to:
New Society Publishers
P.O. Box 189, Gabriola Island, BC V0R 1X0, Canada
(250) 247-9737

LIBRARY AND ARCHIVES CANADA CATALOGUING IN PUBLICATION

Weber, Andreas, 1967-
[Alles fühlt. English]

The biology of wonder: aliveness, feeling, and the metamorphosis of science / Andreas Weber.

Translation of: Alles fühlt.
Includes bibliographical references and index.
Issued in print and electronic formats.
ISBN 978-0-86571-799-2 (paperback).—ISBN 978-1-55092-594-4 (ebook)

1. Life (Biology). 2. Nature. 3. Emotions. 4. Human ecology.
5. Life sciences. I. Title. II Tile: Alles fühlt. English

QH501.W4213 2016   570   C2015-906818-5
                                       C2015-906819-3

New Society Publishers' mission is to publish books that contribute in fundamental ways to building an ecologically sustainable and just society, and to do so with the least possible impact on the environment, in a manner that models this vision. We are committed to doing this not just through education, but through action. The interior pages of our bound books are printed on Forest Stewardship Council®-registered acid-free paper that is 100% post-consumer recycled (100% old growth forest-free), processed chlorine-free, and printed with vegetable-based, low-VOC inks, with covers produced using FSC®-registered stock. New Society also works to reduce its carbon footprint, and purchases carbon offsets based on an annual audit to ensure a carbon neutral footprint. For further information, or to browse our full list of books and purchase securely, visit our website at: www.newsociety.com

MIX
Paper from responsible sources
FSC® C016245

To Emma

Thinking of that early spring day when you had been walking your black poodle and returned so enthusiastically. You told me that you had suddenly seen the flat sandhill with its scattered oak trees behind our house all ablaze with beauty.

The lower animals, like man, manifestly feel pleasure and pain, happiness and misery.

Charles Darwin

All is allegory. Each creature is key to all other creatures.

J. M. Coetzee

The conscious subject is not really perceiving until it recognizes itself as part of what it perceives.

Northrop Frye

Contents

Foreword by David Abram

From Enlightenment to Enlivenment

The book you now hold in your hands is a living thing. Like many living entities, it is intemperate, moody, calm and collected at some moments, filled with passion and exuberance at others. Indeed this book is so brimming with vitality that, if you're not careful, you might find it wriggling free of your grasp and slithering off into the grass. The chapters herein pulse with wonder and are shadowed with ache; the pages are thick with fresh insights and often suffused with a kind of careless beauty....

Andreas Weber is a German biologist and philosopher. This book (a translation and updating of his first general audience book in Germany) is to my mind a necessary text for anyone engaged and excited by the natural sciences. In its pages one glimpses something of the shape that a genuine biology will take if and when our species wakes up to the shuddering ecological predicament that it has wrought. The mounting catastrophes, the accelerating losses of species and habitat, the discombobulation of long-established seasonal cycles as ocean currents go haywire — all these are consequences of modern humankind's strange detachment from the rest of the animate earth. More precisely, they are a consequence of our species' addiction to the countless technologies that regularly insert themselves between our bodies and the breathing land, short-circuiting the ancestral reciprocity between our senses and the sensuous terrain. Enthralled by our own fossil-fuel-driven machines, we forgot our thorough entanglement with innumerable other creatures, with woodlands and wetlands — losing our ancestral attunement to the living land even as the conflagration of fuel steadily thickened the air and defiled the waters.

Weber does not focus on the intensifying calamity; instead he trains his keen attention upon healing the experiential rift that underwrites all this wreckage. For him, this rift is the disastrous dissociation between the thinking mind and the feelingful body. Our material physicality, after all, provides our only access to the earthly world of other animals, plants and microorganisms. It's our sensate, breathing bodies that are so thoroughly intertwined with the soils and the sea currents, with the respiration of frogs, the solar yearning of aspens and the slow weathering of the stones. Since the dawn of modernity, however, we have gotten used to thinking of the mind — our subjective self — as if it were a capacity neatly separable from the physical body. By considering mind as an immaterial essence only contingently related to the body, modern science freed itself to view the physical body as a complicated machine without any inherent sentience, and indeed to view the whole of material nature as a conglomeration of passive objects and mechanically determined processes void of any immanent creativity. Such a stance is intimately related to the more industrial view of nature as a stock of inert or passive 'resources' waiting only to be used by humankind.

Of course, criticizing the mechanical view of nature has in recent decades become something of a cottage industry among the educated set. Similarly, the Cartesian view of mind as a metaphysical substance has lately fallen into disrepute; the search for the material constituents of mind, or consciousness, within the activity of the brain now compels huge numbers of cognitive scientists, neurobiologists and analytic philosophers who regularly gather at immense conferences around the world. Meanwhile the pillaging of earthly nature continues apace; more and more species tumble over the brink of extinction, while more and more ecosystems collapse.

Andreas Weber is doing something different. His richly detailed contribution toward a more mature natural science opens a new and largely unfamiliar way past the thickening mire of callousness, digital distraction and rhetorical smog that clogs so much of our public discourse, confounding even our most private reflections. In this brief forward I can gesture toward only a few of the many themes that interlace in this audacious volume.

Early in the text, Weber discloses a way of resolving the mind-body rift in a manner very different from those who seek the physical correlates of consciousness within the neural circuitry of the brain. For he recognizes that mind, or subjectivity, is less a property of the brain than it is an elemental attribute of the body in its entirety. Any living body, as it navigates its way through the world, must negotiate a host of contingent encounters, obstacles, opportunities and situations whose shifting characteristics could never have been predicted (or programmed-for) in advance. To take a simple example from my own writing: consider a spider weaving its web, and the assumption — still held by many persons of a mechanistic bent — that the behavior of such a small creature is thoroughly 'programmed' in its genes. Without a doubt that spider has received a rich and complex genetic inheritance from its parents and its predecessors. Whatever 'instructions,' however, are enfolded within its genome, those instructions can hardly predict the specifics of the terrain within which the spider may find itself at any particular moment. They could hardly have determined in advance the exact distance between the bouncing branch of an apple tree and the broken bicycle nearby, both of which the spider is using as anchorage points for her current web, nor the exact strength of the blustering wind that's making web-construction rather more difficult this early morning.

And so the spider's genome could not explicitly have commanded the order of every flexion and extension of her various limbs as she weaves this web into its place. However complex are the inherited 'programs,' patterns, or predispositions, they must still be adapted to the immediate situation in which the spider now finds itself. However determinate one's genetic inheritance, it must still, as it were, be woven into the present, an activity that necessarily involves both a receptivity to the specific shapes and textures of that present and a spontaneous creativity in adjusting oneself (and one's inheritance) to those contours.[1]

It's precisely this blend of receptivity and spontaneous creativity by which any organism orients itself within the world (and orients the world around itself) that constitutes the most basic layer of sentience integral to that creature — the inward subjectivity or 'mindedness' necessarily called into being by that particular body as it navigates the present moment.

Weber points out that "intelligence," etymologically, means to 'choose between.' Intelligence, first and foremost, is the ability to make choices, to choose between divergent possibilities. And indeed no organism can live without making choices; even a single-celled amoeba or paramecium must choose between competing routes toward potential food or different ways to avoid a toxin. Such choices may not be conscious in our sense; nonetheless, such decisions stem from each body's felt hunger for more life and from its subjective sense that certain encounters enhance its existence, while others hinder it. As Weber writes:

An organism desires to be, to endure, to be more than it is.
It hungers to unfold itself, to propagate itself, to enlarge itself....
This is a hunger for life. And this hunger is life.

Nor is this inwardly felt subjectivity restricted to animals. Since plants are alive, then their bodies, too, are "instruments of desire" that must improvise their way in the world, orienting themselves and making choices. Weber discusses experiments with plants that are exact clones of one another, carrying identical DNA; even in highly controlled settings wherein room temperature and moisture are the same for all, these plants behave differently, each one choosing a different growth pattern from the others. Each individual plant, it would seem, has its own preferences.

Mind, then, is hardly something generated by the brain. It is, rather, the inwardness necessary to life, the felt subjectivity necessary to sustain oneself, moment by moment, as a living body. This recognition forms the first of three Laws of Desire that Weber proposes as organizing principles for the "poetic ecology" that he envisions. His

First Law of Desire affirms that all living bodies are necessarily bodies of feeling. That the objective structure and behavior of any organism simply cannot be understood apart from the subjective sensations of that organism — that is, cannot be understood without recognizing the organism as a feelingful body motivated by qualitative experiences. All such subjective experiences have their source, according to Weber, in the most basic and exuberant desire to endure and to flourish, which — although it displays itself in wildly different ways — is common to all living bodies.[2]

To resolve in this manner the age-old segregation of the mind from the body is not merely to heal a gaping conceptual wound; it has direct and rather massive consequences for our sensorial experience of the living world around us. If we regard subjectivity as a capacity utterly necessary to the dynamic autonomy of any organism — whether an apple tree, a hummingbird, or a humpback whale — it follows that mind can no longer be construed as an entirely ineffable mystery. To recognize mindedness as nothing other than a body's felt experience of its own dynamic autonomy as it dreams its way through the world is to realize that the mind is not an immaterial spiritual essence that could be housed, sequestered, or hidden away within that being's body or brain. Rather, the mind of any entity is part and parcel of its physicality, and hence is evident and manifest in the dynamism of its material being. The subjective self of any organism, that which feels its encounters and chooses its movements, is not hidden somewhere inside its body; rather, it *is* the body! (Weber's understanding is remarkably akin to what the French phenomenologist Merleau-Ponty wrote of as the subjectivity of the living body itself, which he called the body-subject.) Hence, the mind of a fox is everywhere evident in the manner of its bodily presence, registered not just in the fox's hesitation and the gaze that holds our own, but in the sleekness of its fur, in its way of holding itself, in the abrupt tilt of its head as it catches a fresh scent and the grace of its gait as it leaps the fence and vanishes into the sagebrush. Likewise, the inward, feelingful life of a nightingale discloses itself directly in the syncopated, whirring trills and melodic lilt of that bird's corporeal singing.

For Andreas Weber, in other words, the interiority of a plant or an animal readily reveals itself, to those who have eyes to see, in the outward expression or display of that creature's bodily presence. Here his work echoes that of the Swiss zoologist, Adolf Portmann, who held that the outward surface of any organism cannot help but manifest the innermost self-experience of that organism. This then is Weber's Second Law of Desire, that the wish to live is palpable and visible in the living body of each being. As he unfurls the consequences of this principle, the full radicality of Weber's project becomes evident:

Every organism's inner perspective as a self is simultaneously and necessarily an outward aesthetic reality. If a living being is not an insensate machine but rather is animated by values and meanings, then these qualities become observable. Meaning makes itself manifest in the body. The values that an organism follows are not abstract. They actually guide a body's development and coherence, whether the body is as complex as a human being or as small as a single cell. Feeling is never invisible; it takes shape and manifests as form everywhere in nature. Nature can therefore be viewed as feeling unfurled, a living reality in front of us and amidst us.

By taking time off from our technologies, turning away from the gleaming screens that hold us hypnotized within an almost exclusively human sphere, we begin to loosen our creaturely senses from the over-civilized assumptions that stifle our experience of the breathing world. Slowly, quietly, our skin begins to remember itself to the earthly sensuous. As our ears adjust to the wordless silence, we slowly become aware that other voices are speaking, not in words but in quiet sighs and softly swelling rhythms, in distant howls and nearby trills and cascading arpeggios of sound. We come into the presence of an earth much wider and deeper than our human designs.

And as we turn toward these other voices, tuning our attention to other shapes and rhythms of corporeal experience, our own lives gradually release their knots, becoming more coherent. Following a raptor with our eyes as it swerves and then hovers above a meadow before dropping, talons outstretched, toward some prey we cannot see, or gazing the translucent pulse of three jellyfish as they undulate in a slow rhythm beneath our kayak, we notice a deepening within ourselves, as though a previously unsuspected dimension had just opened within our chest. As though the encounter or meeting with each nonhuman form of bodily sentience — when we recognize it as such, as another shape of subjectivity — immediately confirms, or draws into salience, some essential quality within ourselves. Hence Weber's Third Law of Desire, which he articulates as follows: "Only in the mirror of other life can we understand our own lives. Only in the eyes of the other can we become ourselves. We need the regard of the most unknown. The animal's regard... Only it can unlock the depths in ourselves that otherwise would be sealed forever."

Much of the pleasure of this book consists in watching how the author unpacks these loosely-formulated principles, showing how they play out in richly variegated ways among wildly different species in very different earthly contexts. For a work concerned with the radical interdependence of animate forms and the inextricable wholeness of life — themes that infuse countless well-meaning tomes with a bland and blurry oneness — this book is wonderfully attentive to particularity, divergence and the wild-flourishing multiplicity of the real. If there is any transcendent 'oneness' to be found in this project, it will only be through the immanent, manifold particulars of creaturely encounter in the thick of this teeming world.

Throughout this work Andreas Weber synthesizes a remarkable amount of evidence from a broad array of biological disciplines, forging fresh links between some of the most creative and pioneering researchers in these disciplines — from Weber's own mentor, Francisco Varela and his associates in the field of embodied cognition, to the broad-minded Kalevi Kull and his colleagues in the burgeoning field of biosemiotics, from revolutionary biologists like Lynn Margulis and complexity theorists like Stuart Kaufmann, to plant behaviorists, marine biologists and developmental morphologists. And he draws upon the work of earlier scientists working in the more romantic biological tradition that was once prominent in Central and Eastern Europe, like Johann Wolfgang von Goethe, Karl-Ernst von Baer and the wonderful Jakob von Uexkull, whose crucial contributions were often ignored by the Darwinian mainstream. (My only clear disagreement with this text: Weber's aloof characterization of Darwin's own astonishing work feels unwarranted to me, overlooking as it does Darwin's irrepressible fascination with the diversity and detail of life; neither the regressive views of the 'Social Darwinists' nor the dogmatism of today's neo-Darwinian orthodoxy ought to be associated with the views of the great biologist himself.) Weber elucidates the transformative insights of these many researchers by refracting them through poetry and the arts, but more importantly through his own keenly observed and richly narrated encounters with particular animals, plants and places.

What is at stake in this work is a repudiation of the enlightenment assumption that science is the study of value-free facts, that biology can and must aim at a layer of factual truth, shorn of subjective qualities. For Weber, to strip reality of such qualities is to strip it of life; a natural science given to such a goal can only pave the way for ecological breakdown and collapse. Yet as may already be evident, the participatory ecological ethics that emerges from this work is inseparable from Weber's aesthetics — inseparable, that is, from the renewed attention to the corporeal dimension of feeling and sensorial experience to which this book calls us. Although it would be impossible to consciously pinpoint and assess each of the many subtle relational ingredients of a genuinely healthy ecosystem, Weber suggests that the felt experience of beauty may be our most reliable gauge for recognizing such health. Despite the power of our complex technological instruments and fine-scale monitors, it is the sensate human body that remains — by far — our most advanced, most exquisitely-tuned instrument for feeling into the well-being (or the malaise) of the wider landscape. The sense of wonder, an attentiveness to beauty, the unexpected swelling of joy — these are indispensable guides as we work to bring our communities into alignment and reciprocity with the more-than-human commonwealth. To this end, and in keeping with his critique of the historical enlightenment, Weber's poetic objectivity calls us not toward enlightenment, but enlivenment!

Introduction:
Towards a Poetic Ecology

*Of course the animals need humans. They need us
as if they were old parents, against whom we have revolted
for a while and who one day, weakened, deprived of their
former power, request to be protected by us.*

— Brigitte Kronauer

For 150 years, biology, the "science of life," made no great efforts to answer the question of what life really is. Biologists had a concept they thought to be sufficient for their research: most of them assumed organisms to be tiny machines.

Today this belief is shaken. Only a few years ago we witnessed researchers celebrating the "decoding" of the human genome as a secular breakthrough. They seemed to be on the verge of unraveling the mechanics of life. But not much has happened since then. The boom has come to a standstill. We don't hear much from geneticists these days. Certainly they have been charting the arrangement of genes for a growing number of organisms. But at the next step — understanding exactly how genes make the body and how the body gives rise to feeling and consciousness — the view that life is organized like a chain of military orders fails. In genetic research, developmental biology and brain research scientists are increasingly realizing that they can only understand living beings if they reintroduce a factor into biology that has been thoroughly purged from it for centuries: subjectivity.

Biology, which has made so many efforts to chase emotions from nature since the 19th century, is rediscovering feeling as the foundation of life. Until now researchers, eager to discover the structure and behavior of organisms, had glossed over the problem of an organism's interior reality. Today, however, biologists are learning innumerable new details about how an organism brings forth itself and its experiences, and are trying not only to dissect but to reimagine developmental pathways. They realize that the more technology allows us to study life on a micro-level, the stronger the evidence of life's complexity and intelligence becomes. Organisms are not clocks assembled from discrete, mechanical pieces; rather, they are unities held together by a mighty force: feeling what is good or bad for them.

Biology is joining the physical sciences in a groundbreaking revolution. It is discovering how the individual experiencing self is connected with all life and how this meaningful self must be seen as the basic principle of organic existence. More and more researchers agree: feeling and experience are not human add-ons to an otherwise meaningless biosphere. Rather, selves, meaning and imagination are the guiding principles of ecological functioning. The biosphere is made up of subjects with their idiosyncratic points of view and emotions. Scientists have started to recognize that only when they understand organisms as feeling, emotional, sentient systems that interpret their environments — and not as automatons slavishly obeying stimuli — can they ever expect answers to the great enigmas of life.

These questions include: How does a complete organism develop from an egg cell? How do new biological forms and new species evolve? What distinguishes organisms from machines? Can we design artificial life? What is consciousness? How does the fact that we are living beings structure our thinking and our culture? Why does humanity feel so deeply attracted to nature? What deeper, existential reasons beyond sheer utility impel us to protect nature? In short, what is life and what role do we play in it?

In this book I describe a biology of the feeling self — a biology that has discovered subjective feeling as the fundamental moving force in all life, from the cellular level up to the complexity of the human organism. I also describe how this discovery turns our image of ourselves upside down. We have understood human beings as biological machines that somehow and rather inexplicably entail some subjective "x factor" variously known as mind, spirit or soul. But now biology is discovering subjectivity as a fundamental principle throughout nature. It finds that even the most simple living things — bacterial cells, fertilized eggs, nematodes in tidal flats — act according to values. Organisms value everything they encounter according to its meaning for the further coherence of their embodied self. Even the cell's self-production, the continuous maintenance of a highly structured order, can only be understood if we perceive the cell as an actor that persistently follows a goal.

I call this new viewpoint a "poetic ecology." It is "poetic" because it regards feeling and expression as necessary dimensions of the existential reality of organisms — not as epiphenomena, or as bias of the human observer, or as the ghost in the machine, but as aspects of the reality of living beings we cannot do without. I call it an "ecology" because all life builds on relations and unfolds through mutual transformations. Poetic ecology restores the human to its rightful place within "nature" — without sacrificing the otherness, the strangeness and the nobility of other beings. It can be read as a scientific argument that explains why the deep wonder, the romantic connection and the feeling of being at home in nature are legitimate — and how these experiences help us to develop a new view of life as a creative reality that is based on our profound, first-person observations of ecological relations. Poetic ecology allows us to find our place in the grand whole again. From this vantage point, we can perhaps start to sketch what the sociobiologist Edward O. Wilson has called a "second Enlightenment," no longer putting the human apart from all other living beings.[1]

A fundamental shift is waiting for us. In my last book I called this new logic and worldview "Enlivenment" — the insight that every living being is fundamentally connected to reality through the irreducible experience of being alive. The experience of being alive is not an epiphenomenon, however. It is the center of what defines an organism.[2]

It is still too early to even guess the future implications of this revolution in biology. The neurobiologist David Rudrauf, who works together with the brain researcher Antonio Damasio, asserts that "the search for the way organisms bring forth value and meaning is at the heart of modern cognition research, from robotics to neurosciences."[3] Stated simply, the new biology considers the phenomenon of feeling as the primary explanation not only of consciousness, but of all life processes. By "feeling" I mean the inner experience of meaning — not necessarily from a standpoint comparable to the human psychological reality, but from a certain, individual perspective to which everything that happens is of vital import. Life always has an inside, which is the result of how its matter, its outside is organized. To understand how this inside comes about, in which ways we share it with other beings and what consequences this so far unseen connection has for our view of biology is the topic of this book.

As a science, biology currently finds itself in a situation comparable to that of physics a century ago when basic understandings of matter shifted radically. Compared to the biological mainstream, the new biology is what quantum theory was in relation to Newtonian physics — a breakthrough reconceptualization.

A hundred years ago, quantum theory discovered that observer and observed are not separate entities and that everything is connected to everything else. The new biology I will be exploring in this book adds another, beguiling dimension to our very view of "objectivity." It states that the subjectivity of organisms is a physical factor — an objective reality in its own right. An individual point of view and feelings are not marginal, transient epiphenomenon but rather the opposite: the foundation from which an explanation of life has to start. The new biology places value and feeling at the center of a physics of living organisms — not as one of many interpretive approaches, but as an indispensable element of a scientific description of life.

Biology thus realizes that something identical to our own emotions — something deeply related to our own longing for continuation, our desire to be — qualifies as the epicenter from which the entire spectrum of nature unfolds. This understanding provides us with a home in the wilderness again, in the creative *natura naturans*, that so many people are longing for in their private lives, that they create in their gardens, that they visit during hikes in the wilderness and that they seek to protect. How peculiar and sad that, within the framework of the mainstream sciences, this universal, timeless element of life is seen as a mere curiosity, if it is acknowledged at all.

FEELING THE OTHERS

Nature is not dead. We humans love, seek and long for it. Walking through a forest fills us with peace; gazing onto the ocean calms us. The nightingale's song moves us. We need nature and know we must conserve it. This is self-evident. But at the same time we no longer know if our feelings toward plants and animals are justified at all — or something old-fashioned and rather ridiculous. Feelings and the scientific worldview seem to be irreconcilable. For centuries, many scientists have explained that our joy in other beings is only a sentimental illusion. Such a viewpoint, however, ignores a deep human insight which connects us with other living subjects. Today, researchers are discovering that feeling — the experience of a subjective standpoint — and the desire to exist are phenomena that lie at the heart of a modern concept of biology. This message is so radical that, so far, it is not readily understood. It flouts respectable scientific opinion. Perhaps there is a subliminal resistance to the new biology because it implies a wholesale reconsideration of so many other things. It means nothing less than that the world is not an alien place for humanity, but our home in a profound existential sense. We share it with innumerable other beings that, like us, are full of feeling.

Other beings occupy an enormous space in our imagination. If you ask someone what is beautiful for her, in an overwhelming majority of cases the first answer refers to nature — "a meadow in bloom," "the ocean," "my small urban garden" and so on. We decorate our windows and dining tables with flowers. Fabrics and clothes carry botanical patterns. Stuffed animals lurk in children's rooms. Television broadcasts nature programs in prime time. Urban zoos are always crowded. Many people keep pets. A whole branch of the economy, the tourism industry, generates income by promising access to natural, untouched landscapes.

Humans seek nature because we have lost something inside. In our bodies we are nature. Our essence consists of flesh and blood. We are organic creatures connected by manifold emotional aspects to the more-than-human world, a realm that is not subject to our reasoning alone. Biologists such as the evolutionary scientist Edward O. Wilson believe that mind and feeling have developed in a continuous coevolution with plants and animals — the "biophilia hypothesis."[4] Mind and body have found their forms in such intimate contact with nature that they cannot survive without its presence. Today, many scientists have realized that the fact that we are animals defines our perception in such a fundamental way that we cannot change much about it. We do not experience the world primarily with our minds but with our senses and our bodies — and the consequence of this connection in the flesh is that we perceive the world not as a causal chain reaction but as a vast field of meaning. Human beings think in symbols and metaphors. Mind is meaning as well.

To fully experience this side of our being and to integrate it into our personalities, we are dependent on the presence of nature as a symbolic mirror or a repertoire reflecting or expressing our inner lives. We gather the food for our thoughts and mental concepts from the natural world. We transform plants and animals into emotional/ cognitive symbols according to some of their qualities which are real — or which, at least, we presume to be real. The snake, the rose and the tree, for instance, are powerful organic images that recur in art, myth and cultural rituals throughout human history. These forms of nature seem to have a deep connection to the individual as well as the cultural subconscious. In their living reality and transformations we recognize ourselves.

The poetic ecology that I propose and develop throughout this book connects these deep human and cultural experiences with a scientific understanding of life. Within nature, those values and meanings that life processes naturally produce manifest as living, vibrant forms that are therefore observable by the senses. In the bodies of other living beings existential experiences such as fullness and fear, flourishing and hunger, death and birth are not hidden but visible. They are manifest in their appearance; they are incarnate in the bodies of other organisms. Nature in this fashion exemplifies what we too are. It is the living medium of our emotions and our mental concepts. Given its intimate connection with the formation of our emotional identities, it is no wonder that nature plays such a grand role in human culture. Culture throughout the ages has in many respects been an elaboration of the deep organic connections we share with all other beings—an armamentarium of existential symbolics. One could even say, as Henry Miller once did: "Art teaches nothing except the significance of life."[5] Trees, for example, qualify as symbols for life because in our experience they really *are* life. After a symbolic death in winter they burst into green again. They grow, bloom and bear fruit, without any involvement whatsoever from us. Productivity, adaptation, innovation and harmony, but also decay and failure happen not only to us and our projects, but to all of nature. The power of the elements, the birth, growth and vanishing of other beings, the alternation of light and dark that frames our own inner landscape — the inner and outer dimensions of nature — are one.

But if the longing for nature is a necessary condition of our being, the vanishing of other creatures will have far-reaching consequences. It is possible that in the global environmental crisis, we are about to destroy something without which we are not able to exist. Man may be threatened by an emotional loss that will adversely affect the basic structure of his character. Harvard psychologists believe that by 2020, depression will be the second-most-frequent illness worldwide after heart and circulatory disease, fuelled in large part by a growing alienation from nature.[6] Children in industrialized countries are no longer able to name more than two or three native plants, and adults know more automotive brands than birds' names. In the US, the writer Richard Louv has proposed to add a new disease to the clinical catalogue: nature-deficit disorder (NDD).[7]

But why is nature so important? Because all our qualities — and particularly the most human ones like our need to be in connection, to be perceived as an individual, to be welcomed by other life and give life, in short, our need to love — spring forth from an organic "soil." We are part of a web of meaningful interpenetrations of being that are corporeal and psychologically real at the same time. Humans can only fully comprehend their own inwardness if they understand their existence as cultural beings who are existentially tied to the symbolic processes active inside nature. For humans, the biggest risk of biodiversity loss is that it would bury this understanding. Without the experience of natural beauty, our souls are bound to lose an important part of their ability to grasp what grace means and to act according to that understanding. Without experiencing our real emotional and physical connectedness to the remainder of life, we risk having stunted, deformed identities; we will yearn narcissistically for a completeness we alone cannot achieve. Perhaps the most important psychological role that other beings play is to help us reconcile ourselves with our pain, our inevitable separation as individuals from the remainder of the web of life and our ephemeral existences. The primal feature of nature is that it always rises again, bringing forth new life. Even the most devastating catastrophe gives way over time to green shoots of rebirth and productivity and therefore to hope for ourselves.

THE RETURN OF VALUES INTO NATURE

Many people have objected to the ways in which science disdains our experience of finding our own life embedded in nature's living relationships and consoled by them, regarding such feelings as archaic, naïve or frivolous. Many feel that something is wrong with the reduction of life to a Darwinian struggle of meaningless competition and efficiency. But *what* is wrong could not be perceived as long as the doctrine of a value-free account of life prevailed. For a long time scientists have argued that there is no reality apart from dead matter and that therefore all life must be reduced to the blind laws of survival and selection. This approach defines how mankind is treating the planet. The science-based ideology of efficiency recognizes no values apart from egoistical greed, which it elevates to a law of nature. According to this view, everything else, and particularly feelings such as awe, love and generosity, are viewed as mere tactics invented by our genes for better survival. We tend to banish and ignore that which we know in our hearts is true, and to cling to "facts" that we feel to be false. But as living, physical beings, we always have a compass inside of us guiding us towards what life really is.

We have been perceiving ourselves and the rest of living nature incorrectly because the natural sciences have been studying organisms in the wrong light (or, at least, a seriously incomplete light) for centuries. They have been fixated on understanding them — including, of course, us — as purely physical, external matter buffeted by impersonal forces of nature. In so doing, they have pushed the experience of beauty aside as unworthy of scientific scrutiny, and they have exiled poetic experience and expression. Science deigns to study only "objective knowledge," believing that truth resides solely in the neutral and lifeless building blocks of life. To understand life, we are supposed to join the conspiracy to kill and dissect it. As in a self-fulfilling prophecy, this is exactly what is happening with the biosphere right now. The conceptual framework that we have invented to understand organisms is the deeper reason for our environmental catastrophe. We are extinguishing life because we have blinded ourselves to its actual character. We treat it so cruelly because we believe it to be machinery, raw market fodder, scrap material. But when the Earth is devoid of other creatures, we will be much lonelier. Perhaps then we will realize that we have annihilated a part of ourselves. Along with nature, our feelings are being disabled, perhaps fatally. How we understand the existence of plants and animals will decide our own future, too. This does not mean that we will die of hunger and thirst or that we will psychologically degenerate if there are fewer plants and animals. But we will surely suffer in ways that have yet to be understood. And because body and mind are intertwined in the most intimate ways, because mind represents the body symbolically, in the end it is not only our feelings about life that will be threatened, so will our real lives.

A century of unequaled humanitarian and ecological disasters lies behind us — and without doubt new and even bigger ones lie ahead. How we understand what life is will decide our future. Until now, culture has celebrated a rigid separation of the human dimension from the rest of life. In the last decades, postmodern culture has celebrated this gap as self-evident and denounced any attempts to bridge modernity with "nature" as romanticism or as a misguided nostalgia for authenticity. But this diagnosis is in itself misguided. The real disconnect is not between our human nature and all the other beings; it is between our *image of our nature* and our real nature.

For at least 150 years we have been mourning the disappearance of our soul — and during this same time we have been deliberately sacrificing nonhuman nature on a global scale. These are two sides of the same process. Our task therefore is to overcome this obsessive belief in separation, which has never been the whole truth. The existential imperative for today and tomorrow, therefore, is to rediscover the right balance between our individual needs and the often opposing needs of the whole so that we can flourish. Without calibrating our "ecology of feeling" to the fact that life can only exist as the interpenetration of innumerable lives, the world will truly slip away. We have to learn how we can get back to ourselves by getting closer to "the others" — the living beings with whom we share the condition of "livingness," as Henry Miller put it,[8] the capacity for expressive freedom and creative imagination.

The thoughts set forth in this book inescapably point to a significant ethical choice. We must save nature to allow aliveness to unfold in continuity. Part of this ethic is that we must conserve the presence of other beings for the sake of our own souls. Our own aliveness would shrink without nature or with an impoverished nature. There is a crucial and central place in ourselves that is able to blossom only if connected to the presence of a huge net of other beings and entangled in the give-and-take of those relationships. But this inner center in ourselves at the same time is what points beyond ourselves, beyond the experience of nature as a mere resource for our egos. This inner center is where we are most deeply alive because it is the livingness, aliveness as such, that stirs inside us. It is the creative core of the poetic space we all inhabit, mice and men together. This inner center even precedes the emergence of identity and self. It is nature's center as well as it is the individual's focus. And through it we know that nature is about aliveness. Nature is about beauty because beauty is our way to experience aliveness as inwardness. Beauty is aliveness felt — its potential, its open future, its promises, its tragic possibilities. Nature is the phenomenon of self-producing life making itself visible (and thus of self-producing beauty). It is for this reason that we must save nature. After all, for living beings like us the only meaningful mode of being is to act in order for life to be. We must preserve living beings for life's sake, in order for life to be able to self-organize, to unfold, to experience itself.

In the unfolding new biology, which recognizes feeling as the ground zero of all life processes, our viewpoint must shift towards an ecology of feeling. Only this provides a genuinely new perspective that includes a renewed sense of self and a renewed reason for environmental protection. An ecology of feeling leads us to a new ecological ethics that declares we should conserve nature not because it is useful nor because its complexity has an intrinsic value. We should protect other beings because we love them. We love them because we are a part of them, and even more because they are part of us.

A SCIENCE OF THE HEART

There is a way to move beyond the bleak, lifeless picture of the world that major fields of official science have been painting over the past few centuries. Our perspective can be reversed if the cell is no longer viewed as an autonomic survival machine, but as a being for whom life *means something* and who experiences this meaning as feeling. The revolution in the life sciences thus may generate a truly ecological ethics. This would be an ethics in which the Earth is no longer the neutral stage for an anonymous battle of survival. If nature is the theater in which we experience feelings and develop our identities, then we must protect it because we otherwise would destroy our own selfhood. Only this viewpoint can transcend the void in our current framework of valuing life that cannot explain, by its own philosophical terms, why a *thing* such as a bird or the landscape in which it is nesting and singing, must be conserved. We may intuitively feel that such beings possess an intrinsic value, but it is precisely this value that has been denied and annulled by science as well as by economics.

But the values at stake — the values that current biology cannot explain — are the values of life. They are the values by which organisms create themselves at every instant and by which they organize their experiences. We are able to perceive these values because they are inscribed into our bodies. Certainly not because such a feeling is efficient for survival — quite the opposite: survival is only possible for something that can feel.

This book is directed against the disenchantment of the world produced by the natural sciences and humanities. But at the same time it refrains from proposing a nonrational alternative or substitute for science. Instead, I attempt to explore a third approach: poetic precision. I argue that as living, physical beings interconnected with a living world that is bringing forth existential experience and inwardness, we share a rich common ground with all other living beings. But to reach this new point of departure, we must accept one key premise: this common ground is not objective in the rational-empirical sense. It is defined by *poetic* objectivity. The means that all organisms share an "empirical subjectivity" — a subjectivity that is a defining feature of the biosphere and that manifests as a natural physical force by which they mutually transform one another.[9]

My analysis here will also embrace the sphere of meaning in life — while remaining grounded in the empirical standards of contemporary biological sciences. This goal is all the more important as new nonrational ways of explaining the living world gain popularity. For example, the Intelligent Design theory has been witnessing a major renaissance. This theory, popular in the 18th century with the so-called physical teleologists, tried to reconcile the early results of science with a Christian worldview. It naïvely claims that an intelligent creator invented organisms in the manner of a cosmic watchmaker. Such an attempt does not leave behind mechanical thinking but rather reinforces it, in spite of being a reaction against the strictly mechanistic view of the life sciences. So too, today, with many other nonrational attempts to redeem science from the cold and technical enterprise that it has become — the proposed alternatives say much about our current disenchantment with science but offer little in the way of understanding and enlightenment.

With this book I do not propose a farewell to science, but rather — if you will accept the audacity of the term — a new science of the heart. If we interpret the results of biological research without bias, this is the only pathway that seems possible to me. The biosphere is neither a mechanical structure that has evolved without any sense and meaning, nor a mechanical apparatus designed by an unknown creator. It is alive. And being alive means that it is a constant unfolding of creative imagination that arises from the continuous entanglement of matter and inward experience.

This third way, until now, has rarely been considered: *that matter itself could be creative without a centralized control or planning agency. That matter alone could follow a principle of plenitude and bring forth subjectivity from its very center.* This view is the path taken by poetic ecology and which I intend to pursue in this book's pages. A poetic ecology asserts that life — and not some causal force — is the original animating power of the cosmos. A poetic ecology understands the household of nature less as an economy of checks and balances than as the creative interpenetration of sentient beings.

I have adapted my writing style to support this point of view. I write about nature not only as an object of research but also as a subject of experience — as the place of my subjective experience as a living being. I try to write from inside the living process. Creation can only be grasped by being creative. Imagination can only be echoed by imagining. The question of what life *is* must remain, to be honest and sincere, an unsolved and unsolvable question. To be genuinely understood, the expressive phenomenon of life demands further expression — it must be felt. But the chill, abstract languages of the sciences place a barrier between us and the aboriginal feeling of life. Aliveness remains inaccessible and incomprehensible to "objective" science in the way it defines itself today.

Thus this work is necessarily somewhat personal. I will lead you, reader, into nature and try to make you part of some of my crucial experiences there. But this journey will simultaneously be an expedition into the thinking of modern biology. I will weave the narrative of my own encounters with animals, plants and ecosystems with my analyses, background reflections and reports, explaining science through my experiences. Here I am guided by the conviction that every touch of nature deeply stirs currents of feeling within us, in the same manner as a light breeze stirs the canopy of a tree, the rustle a subtle witness to the atmosphere's restlessness.

PART ONE:
Cells with Aspirations

In this section I describe my great longing for nature when I was young. I was stubbornly convinced that there had to be more in it than perfectly functioning parts. I tell how as a student of the life sciences I despaired over the fact that nearly no biologist could positively explain life, and I will take you with me on the journey to finally discover that we are able to ask the question of what life is, and that some fresh ideas about it greatly change our view of reality. What we will discover sheds a whole new light on biology. It shows that many qualities, which we intimately know in ourselves, and which science for a long time had thought to be existing only in humans, already influence how cells bring themselves forth. A cell does not have consciousness. But feeling, inwardness and value are crucial principles of the phenomenon of life. The drive to selfhood and subjectivity can exert physical powers. How can it be possible that already the simplest being is not a machine? On the next pages I will describe this new sentient physics of the cell.

CHAPTER I:

The Desire for Life

I am the toad. / And carry the diamond.

— Gertrud Kolmar

Have you ever looked into a toad's eyes? They are big and round, like dark waters. But their iris, which centers around the black pupil's opening, has a golden color.

Don't be afraid. Just come closer. Look right into the toad's eyes. Then you will see how these tiny jewels expand into space. Their interior consists of myriads of shiny folds, microscopic canyons and mountains, above which sparkle lost stars. The toad responds to our gaze with the night's sky.

You say that this isn't important? That a toad is an animal and our view of it is uninteresting? Maybe you even find it ugly? Wait a moment. Look again. Look again, as I have done. You'll see that something responds to your gaze. That something exists on the other side, which is no "thing," no mere matter. Something that lives.

I tried for many university semesters to find out what that something is. I tried to find out about life. As a student I dissected frogs in old-style German seminar rooms. I cloned bacteria in crowded labs. I sieved minuscule worms from the muddy flats of the German sea shores to find out about the living. I even graduated with a diploma. But nonetheless I always felt as if something was lacking. As if we, the researchers, the modern scientists, were overlooking something crucial when we observed plants and animals. As if we were blind to what had sparked our interest in living beings in the first place. As if callously disrupting the integrity of a complex apparatus would reveal its secrets and reveal that they are no secrets after all, but "in reality" only links in a mechanical chain reaction.

I have spent my life searching for life's nameless "something." I have finally found it — because science itself has rediscovered it after centuries in which researchers were determined to ignore what they could have seen all the time, were it not for their self-imposed duty to be objective. Maybe I have also found it again because I always wanted my feelings to correspond to my thoughts. Everything started many years ago with the toad. Or, to be more precise, it started with a whole pond full of amphibians, full of diverse species.

It was one of those silvery gray days at the end of the winter in northern Germany. Brown, bristly hedges framed the small road wending through the bare fields that lay motionless under the transparent air. It was so cold that the wind hurt my ears as I rushed along on my bicycle. But already a certain warmth had settled in the atmosphere in all the gray colors that had subsisted from the long winter, some ineffable promise that had drawn me out into nature.

I was 14 then. School was out and I was pedaling through the fields to one of the small ponds that was hidden behind a sparse path of reeds, in the broken shadow of elm and beech trees and hazelnut bushes. Some of these were not much bigger than puddles in which the winter's moisture had gathered and now hesitantly started to sparkle. In its small ponds the country opened its eyes to the sky. I went to one modest body of water that often was the destination of my solitary bike trips. I had been here with a couple of friends when it had still been winter. Mara Simon was among them, a tall blond girl with an ironic wit who barely talked to me. Only a few weeks ago we had been skating here. Then, the pond was covered with a massive layer of transparent ice. I had been watching the water weeds beneath it like in an enchanted aquarium, a world silent and in peace. Walking home through a quickly falling dusk, we had been telling each other the scariest horror stories we could remember. I was hopelessly in love with Mara, but she did not notice me. When she had been dashing over the ice I had tried to outrun her. Every time I caught up she threw back her head with its blond pageboy haircut and flung me a scornful smile with which I knew I had to content myself for the coming days.

Now, some weeks later, the frozen surface had completely melted without a trace of ice. The pond had completely changed its aspect. I let my bike drop into the withered grass and bowed down to the water. When I looked closely I could see that its smoothness was not complete, its mirror not perfect. The black marble of the surface cracked again and again. It sprang open and formed bubbles like under an endlessly slowed-down rain. The surface crackled and sent tiny waves to the shore where I was kneeling. It seemed as if some mysterious effervescence sparkled out, some enigmatic ingredient of life that I did not understand.

I lay down on my belly and crept closer to the water. When I pushed my arms forward beyond the muddy waterline the liquid enveloped my wrists with icy force. An abrupt movement inside the pond made me pause. Mud rose from the bottom and slowly sank back. I stared through the water.

Then I saw the animal. Nearly unnoticeably, it took off from the muddy bottom and went up in the water like a slender elongated balloon, stabilizing its course with its four plump legs and straight tail. With its snout it pierced the surface and sent a flat wave over the pond. It gasped for air — and sank back. I saw that its lower side shone bright orange. A serrated comb was waving on its back and on its tail, like on a fairy-tale dragon.

My heart leapt. The depth of nature had suddenly opened, here on an unadventurous school day in late winter, in the twilight of the northern German plains. The creature before me silently drifted through the water, rising so slowly as to be nearly standing still. I discovered a second animal, than another one. They came upward and inhaled some air, then fell back into their floating state. Through the water I saw their eyes, which gazed through me but which nonetheless sucked me into them. In these magic crystal balls, life was trembling. The longer I stared into the liquid space before me the more strongly it expanded until finally it seemed to me that in the silent stare of the animals, the water itself was looking at me.

After a long while I got myself together, stood up and cut myself loose from the animals' gaze. Some days later I bought an amphibian guidebook. The animals were newts, *Triturus vulgaris*, and the dragon's comb on their backs belonged to their spring wedding dress. I could not know then that I would never see them here again.

INWARDNESS WE CAN TOUCH

I was a child then. Or rather, I was at the age when I was supposed to slowly morph out of my childhood. But I was slow. The others changed drastically. They started to ask their parents for small mopeds. Soon I was meeting them with their scraggly arms and helmets that seemed far too big for their meager bodies as they darted along on their brand new Zündapps and Hondas. They squatted on their machines in front of the schoolyard, together with some cooler and older dudes. They smoked. It was a group to which I had no access. Mara Simon was now one of them.

For a while, I was growing out of childhood but not into adult life. I grew into nature. I was 15 and had rented a tiny garden close to the housing development that our narrow townhouse was part of. It was a triangular patch of earth between a sandy parking lot and a barleyfield. In the afternoon, when the others met to have a smoke and chill out, I was kneeling in my vegetable beds, sowing lettuce and borage and pulling out couch grass or goutweed between young carrots.

Nonetheless, my path did cross Mara Simon's one more time. It was a sunny Wednesday afternoon. I was pushing a wheelbarrow along the road. It was loaded full of pig manure for my garden, which a local farmer had sold me for the absurdly expensive sum of 50 deutsche mark. I had spent all my savings on it. Mara appeared on her bike when I crossed the street, her whitish blond hair swaying in the wind, her cheeks red and radiant. I quickly looked down at myself: worn-out corduroy trousers, boots smudged with manure. I held my breath. When Mara zoomed past I tried to capture her eyes with a helpless look. But she looked past me. Her gaze was as closed as if made of bullet-proof glass. In the evening I wrote her a letter. She responded with a red felt pen. "For me other boys have the right of way," it read, politely implying some relation between my vehicle fully loaded with dung and the Zündapp mopeds of her usual companions.

I often pedaled to my pond in the height of this summer, not knowing it was the last one I would spend there. When the water-weeds and algae grew, the newts began to expertly hide themselves. I could not spot them any more. But I'd discovered a couple of fat common toads and marvelled at how the wiry males in the water embraced the females, who were nearly bursting with eggs. The males emitted such an air of comfort on their fleshy rafts. I listened to the brown frogs' low growling. I sat still at the shore and watched how their golden eyes reflected the sun. Just before winter I moved away with my family. I never saw the small pond again.

I had found a home here for one summer season, a home among the animals, with their simple needs and their forthright expressions of these needs. I had a home here, but what it offered was not forgetful salvation. It gave me some wordless understanding of the bigger drama I was part of. It did not console me about Mara, but it did situate that emotion within a larger constellation of feelings. I started to learn then that nature is not a place that shields us from feeling; rather, it is a refuge where we can experience our true emotions. Plants and animals help us discover significant things about ourselves. In them, we find our own inwardness.

With an indulgent laugh, we have always allowed dreamers and poets to experience nature in this way. But we never took it too seriously. Since my own adolescent experiences, however, I have encountered many serious people — not just romantics and dreamy children — who regard such feelings as the center of gravity for life itself. Science itself has come to the paradoxical realization that it can no longer reject emotions and values if it wants to understand life. In this new light, I can now explain my experiences from those distant summer days. But it has taken some time.

For centuries, philosophers and scientists have been working in the opposite direction. Historically, science's goal had been to decode the cosmos made by God, to discover the universal laws which keep the particles of matter in their appointed circuits, like the idea of gravity explained the stars' trajectories. The first biologists in the 18th and 19th centuries zealously approached nature as I did in that year at the pond: enthusiastically lusting for novelty, for fresh impressions, for exuberance. Within a few decades, those early scientists discovered and described hundreds of thousands of species, classified dozens of habitats and gave birth to the sciences of morphology, biochemistry and physiology.

In this first biological research boom, physical science established the benchmark standards according to which the young science of biology was to be measured. The living world was to be experienced with all senses. Empirical inquiry required nothing less. But the conceptual framework presupposed a mechanical system. The French philosopher René Descartes had declared that living beings were nothing but more or less complicated machines, and so the task of the naturalists swarming out into nature was simply to make an inventory of the sheer diversity of these mechanical devices. Still, many biologists did not give up what they felt to be the crucial questions of life.

These were the very questions that had lured me into the outdoors that faraway spring and summer. I, too, wondered: What distinguishes animate from inanimate matter? What transforms the billions of particles and hundreds of thousands of chemical bonds into an organism with such a complicated form — and why exactly one like this? Why is a baby, at only five months, able to discriminate between a machine and a living being without even a moment of hesitation? That summer I lay at the bank of my pond and brooded over enigmas that for a long time had been the central problems of biology.

This fascination never left me. When, a couple of years later, I started my science studies, I was confronted with a rude realization: biology claimed it had already found answers to all these questions. But its answers looked totally different from what I had imagined.

Biologists had clearly borrowed the conceptual models developed by the physical sciences. The primary tenet for understanding life was to study the component parts of dead matter that comprised living machines. After the synthesis of evolutionary theory and genetics beginning in the 1940s, biologists completely disregarded the importance of feeling and the experience of inwardness. For them these dimensions were mere subjective illusions and sentimental reminiscences whose purpose could be easily explained by a Darwinian calculus: feelings are obviously nature's way of inducing individuals to behave in ways that generate the most possible offsprings carrying one's own genes. Any other picture was considered unscientific, and for many, still is. To call scientists' work unscientific in public is the ultimate stigma in our society, not totally unlike accusing their work of being "bourgeois" in the 1950s Soviet Union, earning them a one-way ticket to the gulag.

Most biologists, I came to understand during my undergraduate studies, still try to comprehend organisms within the conceptual framework used by physicists to explain the universe in Sir Isaac Newton's time — as a system built of separate entities that each predictably respond to external forces and obey eternal, unchanging laws. The "golden age" of the life sciences was entirely focused on discovering the biological counterparts of these eternal physical laws. Charles Darwin discovered the law of selection, which affects organisms in the same way as gravity affects planets. Gregor Mendel discovered the law of heredity, the biological equivalent to the law of the conservation of mass. James Watson and Francis Crick found the genetic code, the meta-law that ordains how specific organisms are constructed. In principle, it seems, these laws are thought to provide everything that is needed for the universe to generate a living being — and their careful application therefore in principle can lead us to generate organisms from scratch (as synthetic biologists are now attempting).

Biology in the course of the 20th century became a prima donna among the sciences. It offered, at least in principle, the knowledge necessary to create life. At the moment, this pretension has shifted to the field of synthetic biology, with its massive allocation of academic research power to the material reconstruction of life forms and their functions, and also with its potential of doing hip stuff and creating unexpected novel life forms in makeshift kitchen labs with DIY enzyme kits. The science of life has superseded physics as the most exciting research frontier. This shift is also due to the fact that physics, after centuries of technological success, has been stuck in a research morass, trying to sort through the many unresolved enigmas left by Albert Einstein, Niels Bohr and Werner Heisenberg, among others. Physics stopped being only the mathematical description of a causal lawful universe and became complicated, difficult to understand, but utterly interesting. Biology, however, just worked. In general relativity theory, space, time and observers are entangled, and there are no objective points of reference any more. In quantum theory even the spatial and temporal separations of events cease to be. Everything that is, was and will be, is connected to everything else. In biology, however, organisms have become as predictable as atoms had been in an outdated physics, guided by eternal and external laws.

Even though biology has triumphed as the preeminent science of the second half of the 20th and the first decades of the 21st centuries, it has not yet hosted any really important theoretical or empirical revolutions, as the physical sciences did. That is not to deny that biologists have amassed a mountain of knowledge enabling them to probe deeply into the inner functioning of organisms. With startling speed, researchers have now decoded genomes for everything from *E. coli* and barley to mice and mammoths to man. Unfortunately, they are still imprisoned in the Newtonian frame of mind. In the course of their translation frenzy, geneticists have had to admit that the simple linearity of the one gene, one function model, which Watson and Crick proposed, is not how DNA works. Genes mostly do not have only one function, but several, which themselves depend on many external factors and the general condition of the cell. This has meant that no disease can be successfully eliminated by genetic means alone.

At the same time, biological explanations for human behavior that view feelings and actions as results of supposed genetic behavioral modules remain very much disputed. And the theories that sociobiology has proposed to explain our behavior — for instance, that a woman is averse to adopting the children of strangers because they do not carry her genes — are at best ad hoc and far-fetched speculations.

Biology, still captive to the Newtonian model of physics, has been treading water. In some respects its situation resembles that of physics a century ago, when it declared with smug certainty that all of the big problems about matter had been solved, leaving only minor calculations to be done.

Recently, though, everything has changed. Frustrated by the inability of traditionalists to answer many large, open questions, some daring biologists have started to renounce the central dogmas of their science. They not only want to find greater personal creative space, they recognize that more and more novel empirical findings in biology simply cannot be explained by mainstream teachings.

The late American biologist Lynn Margulis, for instance, proved that the great diversity of shapes and forms of all higher organisms' cells had not come about by fierce competition and natural selection but by the symbiotic fusion of simpler precursors. Biosemioticians like the Estonian Kalevi Kull, the Dane Jesper Hoffmeyer and the Italian Marcello Barbieri have introduced a particular version of relativity into biology. They no longer view DNA as a machine code to execute the binary orders of a genetic blueprint but rather as a musical score that the cell can orchestrate in many ways according to its momentary needs.

Like the revolution in modern physics, such a perspective abandons absolute causality. Cells react to external stimuli but also to the triggers that come from their own metabolism, with a certain autonomy. How they behave depends on their inner state — and also on what an observer does or does not do. It turns out that even micro-organisms are highly complex and intelligent, not simple and mechanical. They are certainly not ticking molecular clocks whose genetic software acts as a kind of remote control. Experiments have long shown that body features such as fur coats and leaf surfaces self-organize without needing any genetic instructions. Biological order arises "for free," as the complexity researcher Stuart Kauffman has famously stated.[1]

This idea is widely confirmed by observations from developmental genetics. Certain embryonic growth centers unfold somewhat independently of DNA "instructions." Once set into action, observe developmental researchers Gerard Odell and Gerald von Dassow, an embryonic growth center does not obey genetic orders but rather interacts with them, transforming certain impulses in ongoing development according to its particular situational needs. Embryonic regulation circuits, the authors hold, are "robust" in respect to their physical and genetic environment.[2] In the new perspective these scientists are beginning to establish, cells appear to be units of will, purposeful agents. Their most important feature is the fact that they consistently renew themselves and bring forth all the parts they consist of. They literally create themselves. Cells show a breathtaking perseverance. They really will do anything to ensure their continued existence. In their steady exchange of matter with their surroundings they do not resemble inert machines in any way. They spiral upward in a continuous dialogue with the hereditary material, but they are not directed or governed by their genes. DNA is a scaffold for the flesh, not its blueprint. The body must "read" the genes. It must make sense of them, interpret them and integrate them into its own logic of self-maintenance, as Harvard University developmental biologists Marc Kirschner and John Gerhart observe.[3]

THE LAWS OF DESIRE

It was this autonomy that had given me so much deep pleasure that summer, alone at my pond. Completely oblivious, the animals floated in their water. They were a part of it and still so markedly unique and self-sufficient. Their clumsy rising and sinking, their grotesque dragon-like appearance — the newts were worlds in themselves. They were independent centers of life. They made me understand what Aleksandr Solzhenitsyn might have had in mind when he wrote in *The Gulag Archipelago*: "Every being is a center of the universe."

If today, at the beginning of the 21st century, biology recognizes this autonomy, it has taken a century-long detour before stumbling over something quite obvious. For already in antiquity, the philosopher Aristotle, himself a famous animal researcher, believed that living matter at its deepest core is engaged in a profound struggle for being. This striving, this yearning for its own existence to persist, is its soul, he said. Maybe Aristotle had touched upon something quite fundamental. An organism desires to be, to endure, to be something other than what it is. It hungers to unfold itself, to propagate itself, to enlarge itself, to suck in more of the precious stuff of life — that you can possess only when you are breathing. This hunger is life. The desire to live carries an organism like a wave carries a swimmer in the ocean.

When I think back to those days, to that afternoon at the quiet pond, I know that my behavior was absolutely scientific. I was keenly observant. I carefully surveyed the water and discovered a law of nature, without actually realizing it. In looking at the newts, I experienced the force of life in a completely direct way. I experienced the same force that holds the cell together, which defines its urge to continue to exist — the same force that makes me move and carry on.

Today I can give a name to that urge. Today I know that it is the centerpiece of a new science of life. This new science has progressed beyond the dogma that its truth must always be measurable, objective, neutral and value-free. This is no longer possible, according to the latest results of complexity research, systems biology, biosemiotics, robotics and brain research, which I will describe in the following chapters. Increasingly, it appears that living beings are actually quite different from what they have been supposed to be in mainstream biology's perspective. Life is nothing mechanical and neutral. Life is value that has taken physical form. It is deeply marked by the experience of value because for an organism to exist does not mean that it is merely "there" as matter is, but that it is always building itself up in order to be. There is a profound interest at work which values being alive over dying. I will come back to this new idea of an intrinsic felt value or self-interest as the center of biological reality throughout this book in a variety of ways.

If we push open the doors whose keys have been provided by researchers like Kalevi Kull, Jesper Hoffmeyer, Lynn Margulis and Francisco Varela, it becomes clear that a transformed biology must use a new vocabulary with its new style of thought. It has to speak about life in a way that acknowledges both its physical and subjective sides and balances them. The new language must do justice to behavior — but also to the feeling that the behavior and its experienced aspects, *mean* for the organism as a subject. Within the framework of classical biological science, this may sound like heresy — and it poses serious methodological challenges, to be sure. But whoever dares to breach the entrenched biological dogmas is rewarded with an inestimable gain: nature and mind are no longer separated. We can glimpse the actual functioning of organisms. The deep chasm yawning between ourselves and other beings, between how we feel and how we describe the world, closes for the first time. For the first time, we are welcome. We belong to the world.

In this new biology, the drive to live — which we can observe everywhere and which we also know from within ourselves — becomes the primary axiom. It replaces the blind "struggle for life" of biological Darwinism. To make clear some of the unexpected consequences of my viewpoint, I will call this striving to exist the *First Law of Desire*: everything that lives wants more of life. Organisms are beings whose own existence *means* something to them. They have interests like us — although these interests are not necessarily conscious. But all organisms are carried by the same sensation that human subjects feel from the very beginning: existential concern with the meaning of every encounter, the touch of every other being and the things that other beings produce. Organisms are focused on being in the right place and on avoiding the wrong place — the way the rose beetle senses that the inside of a rose is a good place for him and the paved road a bad one.

What completely mesmerized me in that long past summer was the fact that I could *see* that concern. It radiated out in every tiny movement, in every gesture. It became clear that even if plants and animals do not have consciousness, they nevertheless have an *inside* with a distinctive, vibrating gestalt. I could nearly walk into it. I could dive inside it and touch it. I saw that the inner life of organisms displays itself at their surface. Like all important insights in nature it is not hidden, as Heraclitus would have us believe; it is revealed.

Therefore we have to posit a *Second Law of Desire*. This law subverts the dogma that all visible characteristics of an organism must finally serve the selfish mission of reproducing as many healthy offspring with its superior genes as possible. According to the Second Law of Desire, the power of the wish to live is palpable — and never invisible. It is always present in the living body. The desire for life springs from hunger and thirst, and replenishes itself in growth and joy and thus becomes transparent in the body's gestures. Organisms therefore are not one-track machines. They are instruments of desire.

On that afternoon beside the pond I experienced the opposite of what science has believed since the Renaissance — that the mystery of life can be cracked only by physically dissecting life, by "torture on the on the rack of experiment," as Goethe put it. No. There is no need to tear apart anything. The enigma is hidden in plain sight. The depth of life shows itself at the surface. Think of the fireworks of autumn foliage, think of the weightlessness of jellyfish in the ocean, which expresses that organism's signature, free-floating way of being, its peculiar and constant intercourse with the perfumes and petals of the ocean. Life generously discloses its laws, the conditions by which it can prosper or fail. It reveals itself everywhere. We need only allow ourselves to see and to touch to understand life on its own terms.

It was this understanding that offered me a sort of vague, but joyful insight that summer at the pond. It was not human soul or human consciousness that I encountered among the silent amphibians of my little world at the edge of the water. Life itself was making itself transparent and embraced me. I beheld the feelings of life, which had taken a sensual shape — a body — and I could find my own feelings in these shapes.

For that reason we must introduce a *Third Law of Desire*. This law rejects the idea that the only things we can truly understand are of human make and that all other beings must remain endlessly foreign to us, as philosopher Thomas Nagel seemed to point to in his famous essay "What is it Like to be a Bat?" in which he declared all attempts to intuitively connect to other beings as futile. The Third Law of Desire, however, holds the opposite. It emphasizes that we must be close to other living beings in order to grasp certain depths of our own nature, certain ways of our transforming power and our deeply imaginative and creative character. The Third Law of Desire states: Only in the mirror of other life can we understand our own lives. Only in the eyes of the other can we become ourselves. We need the real presence of the most unknown: the owl's mute regard, the silent newt's gaze. Only *it* can unlock the depths in ourselves that otherwise would be sealed forever. We need the experience of an inside unfolding in front of us, displaying itself as a fragile body. We need other organisms because they are what we are, but with this cunning twist: they are that hidden part of us which we cannot see because we exist *through* this part and we see *with* it. Viewed in this light, other beings are the blind spot of our self-understanding, its invisible center, which is the source of all vision.

These reflections about our need for the presence of other beings bring us to one of the biggest dramas of our epoch. Through the gaze by which nonhuman beings look at us, they are helping us forge our own identities. We are entangled with others; why this is so and the ways in which these counterparts are indispensable to our imagination, are what I want to show during the course of this book. Other beings measure us with eyes that are gracious but also demanding. They give, but they expect an answer, too. The silent gaze of the newt is a question directed to us. Our response, too often, is to avert our gaze in denial.

At no other time in history have more species been disappearing from Earth than today. Amphibians, of all animals, are the most devastated class. In the late 1980s researchers started to observe that more and more populations of newts, toads and frogs all over the world were suddenly dying. They did not trickle away slowly during years of increasing infirmity; they just vanished from one year to the next. Until the late 1980s, thousands of golden toads made a rainforest swamp in Costa Rica boil every summer. Some years later only one male was found by a scientist, and then no more at all. The golden toad, only discovered in 1964, is now extinct. This drama of loss after loss in the amphibian kingdom continues today. In the last few years alone, more than one hundred species have vanished, and with them, countless unique expressions of the ineffable embodied in the tissue of life.

Scientists continue to debate causes. Some think that increased ultraviolet radiation could be responsible. Others blame alien fungal infections. An American toxicologist has claimed that Roundup, the infamous Monsanto herbicide that continues to be sprayed on crops in nearly every corner of the globe, is involved.[4] But by far the most obvious reason is that less and less habitat for these creatures remains. "Development" has wiped out countless small bodies of water in which the Earth looks at itself through dreamy eyes, like my small pond seemed to do. The amphibians have vanished, as if they had resolved to boycott this world for all time to come.

A species dying out makes no sound. It vanishes discreetly. It goes in silence as if it did not want to create an uproar, as if it did not want to disturb us any longer. We do not notice unless we look carefully enough, unless we search every springtime for the beings we have taken for granted throughout our lives. We realize their absence only if we wait for them in vain, if we return day after day, and the anticipation slowly turns to hollow grief.

31

On one of those summer days I collected a handful of frog's spawn. I drove my fingers into a lump of eggs floating in the water to split off a part from the whole. It was nearly impossible. The mass stubbornly stuck together. I could not see where one egg finished and another began. I did not want to destroy any of them. Although the lump of spawn was divided into many units, it nonetheless appeared as one whole thing. The resistance I encountered was vigorous and compliant at once. I had a feeling of pleasure and of inhibition at the same time. It was the kind of shudder you might feel if you were to bite living flesh.

Later I put my lump of frog life in an old aquarium. To observe the development of amphibian eggs is to experience a kind of magic — the transformation of a unidimensional dot into the vigor of life. The "it" becomes a "thou." Nearly a nothing, it starts to pulsate and to unfold into a being, a universe unto itself. The black pinpoints in the middle of transparent glass bodies elongate, contort, inflate and finally thrash into freedom. Life is opening its eyes through a hundred quivering, darting pupils.

I had not put a cover on the jar. As long as the tadpoles wriggled through the water without limbs it had not seemed necessary. But then they grew delicate hind legs, and shortly thereafter the frail forefeet started to show. Following a hot early summer day all the frogs were gone. While I was away, they had made a particularly big developmental leap and hopped out of the aquarium. Weeks later when I cleared up my room I found them behind the heater and under the wardrobe — hardened mummies, frozen in the gesture of their frogness. It was a small goodbye heralding a much larger one.

In the animals our inwardness stands before us in an unknown shape. If we lose them, we do not just lose precious, fascinating creatures. We lose ourselves. We renounce something profound about our condition of being in the world. We forsake ways of being creative, ways of giving birth. Each species we lose today is the permanent loss of a manner of expression of a living cosmos. After it has gone, reality will never be able to express the same gesture. Each species has become a singular expression of livingness through the fortunes of time and, as the American naturalist Williame Beebe put it, could be replaced only through a new evolution, through a new Earth history.

Species are unique creative imaginations, variations on the theme of being which no instrumental consciousness could have imagined. The female strawberry poison-dart frog incubates her eggs far from water and then places the freshly hatched tadpoles individually in tiny ponds, which have formed on large leaves. Every few days shereturns to these ponds and feeds the young with unfertilized extra eggs she produces. The spawn of the common midwife toad comes in long rows of single eggs looking like beads on a string. The male adult winds the strings around his hind legs in order to watch over them, and to carry them for a swim if they turn dry, until it is time for the eggs to hatch, at which moment the father looks for an adequate body of calm water and deposits the spawn.

It is the presence of other beings, the gift of their being here, bestowed upon us and not created through our powers, that grants inexhaustibility — and through this, hope. Their presence is that which does not comply with human measure and which does not obey our power. It adamantly ignores human logics and planning. Their presence is a miracle — a free treasure whose value we can barely imagine. An undeserved, unearned gift that falls from above. An embodiment of grace.

THE ANIMALS' GIFT

A couple of years later, during my studies in the southern German city of Freiburg, I helped toads safely cross a road in the forest for several nights. It was a mild February, and a fine web of mist often filled the air. I was alone. I was relieved to be able to walk around in the dark where nobody could see me. I was fond of letting myself sink into the blackness of the night. A few days earlier, a friend of mine had raced her car into a wall at 60 miles per hour, apparently on purpose; there was no evidence that she had attempted to brake.

To help the frogs cross the road, we had stretched a low mesh-work of wire along an outbound highway. At a few yards distance, buckets were sunk into the ground, into which toads were sup-posed to fall as they hopped along the fence in search of a hole to pass through. Nothing happened until eleven o'clock. The buckets remained empty and I walked aimlessly. Then suddenly there were animals everywhere: toads, brown frogs, agile frogs and other species — alone, in tightly paired couples and even in clusters of three or more, holding one another closely. They plopped into the buckets, which overflowed within a few moments, and spilled over on to the ground as if a gigantic faucet was filling a watering can. The animals covered the ground as if they had fallen from the sky. They crept over my feet. Their eyes shone in the light of my lamp as they wriggled from between my fingers, trying to escape my grasp.

For hours I labored in an environment of living flesh. I picked up animals, poured small buckets into larger ones and dragged the containers filled with croaking, cooing creatures across the road. As late as the 19th century, scholars had suggested that frogs and mice could arise *de novo*, spontaneously materializing from rotting straw or damp rugs. How naïve. And yet this late night phantasmagoria of pulsating flesh seemed a creation out of nothingness. The world had stirred and abruptly given birth to "supernumerous being," in the words of poet Rainer Maria Rilke. It sprang forth from nowhere into the beam of my flashlight. In the bushes an invisible robin sang and filled me with gratitude.

My silent exchanges with the teeming throngs of amphibians these nights soothed my sadness. I accepted the animals' gift and tried for a few hours, tired and silent, to give them something back, to escort them to safety

Many years after my idyll at the small pond, I made the mistake of returning. It was some school anniversary, and I unexpectedly ran into Mara Simon in the evening. The skin around her eyes showed tiny wrinkles. She seemed captivated by my small talk, which had become more ingratiating and practiced following a few lucky conquests. For a while I had Mara all to myself. Her gaze was rapt, her eyes glistened. It was our first conversation at eye level, and I finally had her complete attention if not her surrender. But I suddenly did not care.

The pond was gone. It was now a turnaround for tractors, the mud rutted by deep tracks. The jagged calls of crows pierced the defenseless air.

CHAPTER 2:

The Machine That Can Die

We hang here, inquisitive carbon-based life forms,
knowing that every atom of carbon now in our
bodies was once in the interior of a star.

— James Hamilton-Paterson

The rental car was stuck up to its doors in mud. With every push on the gas pedal, fountains of dirt squirted from the wheels, and the Ford Focus tilted a little closer towards the ditch. Between attempts to get the car free I watched Riste as she gazed through the spattered windshield into the forest. The pale stems of birches were fading into the twilight, and spruce trees slowly became stark black silhouettes.

Everything was vanishing into the diffuse shimmer of early evening as if the landscape had changed suddenly into a negative image of itself, as if the inside of the world was gently opening before us. A cloak of gray snow and dull birch bark had embraced us from under which no escape seemed to be possible.

The luminescent numerals in the dashboard showed 5:30 p.m. The windows were steaming up. When I had decided to turn the car, the grass at the side of the dirt lane had only a dusting of snow and seemed solid. But the weight of the Focus was too much, and it immediately sank into the grass as if it was a quaking bog.

I shifted into the third gear and firmly pressed down the throttle. The engine let out a high-pitched roar and the tires sprayed mud everywhere. Riste, looking impassively ahead, said, "This is normal here when spring arrives. We'll have to walk to the house." When I did not react she placed her warm hand on my fingers clutching the gearshift. "We can make it there on foot," she said. "I know a shortcut through the forest." Her eyes sparkled. Outside, the vast expanse of the forest blurred into the dusk.

It was late winter in Estonia. I had flown to the small Baltic country to meet with a researcher and philosopher at the country's famous university in Tartu, who could teach me what an organism really is. Kalevi Kull is part of a growing vanguard of biologists working on a new view of life. The Estonian is among the founders of a movement called "biosemiotics" — the theory of nature as a sphere of meaning and experience. Biosemioticians view nature as a realm not only of cause and effect but also of interpretation and signification. Biosemioticians consider feeling and value to be the foundations of all life processes. They start with the results of cutting-edge biological research while recovering the nearly forgotten ideas of brilliant holistic thinkers of the past. Biosemiotics is one of the most important tributaries feeding the wide, robust river of holistic biology, which I will explore further in this chapter. It seeks to consider both body and inner experience as biological phenomena. It takes account of the paradoxical fact that while organisms may consist of nothing but matter, they nonetheless exhibit many surprising, unexpected qualities such as the experience of subjectivity and inwardness, or *self*. Biosemiotics is the attempt to bring these two poles together as two aspects of a unified reality. It aims to overcome the split between "material" body and "felt" experience that has for so long obstructed our understanding of our own lived reality.

A native Estonian, Kalevi Kull sees himself — with a dash of patriotic pride — as the inheritor of a long tradition. The German-speaking areas of Central and East Europe have long hosted a fertile and prominent school of romantic biology. Its arguments concerning the philosophy of organisms were in the forefront of scientific thinking for a while, claiming the empirical and expressive modes of perception were linked. This claim is personified in one of the first founders of a poetic natural science, the writer and researcher Johann Wolfgang von Goethe. An eminent biologist following in Goethe's tracks was Karl Ernst von Baer, also teaching in Tartu (then called Dorpat), Estonia. And even the Prussian philosopher Immanuel Kant was somewhat under this influence. Kant developed some groundbreaking ideas about organisms in his later texts that still influence the holistic tradition in the life sciences today. Kant even created the term "self-organization." It is not unreasonable to think that the vast, scarcely touched wilderness zones of the northeast in some way furthered the development of these ideas. Nature was a raw, powerful force; it was real — and beautiful. Some of the most significant thinkers in holistic biology have come from this untamed region, the area of modern-day Estonia, so it is fitting that this wildly romantic Baltic region would continue, in the 21st century to germinate the seeds for a view of life that never quite took root the first time around.

The hardiest of these seeds is the idea that causal-mechanic biochemical laws alone are not enough to clarify a living being's character; its uniqueness cannot be explained by a mere description of its externally observable behavior, however exhaustive. Biologists such as Kull believe that the question of what "binds a living being's innermost core together" — as Goethe's Dr. Faust put it in his famous opening monologue — is as poorly understood today as it was in Goethe's time. We are still searching for a formula to explain the phenomena of experience and matter and how they are entangled so mysteriously with one another. We have yet to formulate a "poetic materialism", a worldview that accepts the materiality we and the world are made of, but at the same time admits that this inevitably leads to a poetic ecology of meaningful relations and inwardness.

I came to Estonia seeking answers to this old, painfully unresolved question. After many years of searching for the gates to this kingdom, I had at last discovered the ideas of biosemioticians and the work of other contemporary holistic biologists who were seeking to dethrone the dogmas that had ruled the life sciences for so long. The new theories developed by researchers like Kull finally provided a biological foundation for the three laws of desire I introduced in the previous chapter, making a poetic materialism possible if not yet entirely persuasive. These biologists are today laying the foundation for a radically different understanding of life. They believe that any account of a living thing that looks only at its organization, shape or behavior cannot be adequate. An organism whose inwardness is ignored will remain shrouded in mystery.

A WORLD PERVADED WITH LIFE

The poetic ecology to which philosophers like Kull are contributing is not more complicated than the mainstream form of life sciences. But it is new, and hence it can be difficult to grasp it quickly and intuitively. At first I had trouble understanding its underlying ideas. And then at some point I could not think in any other way. My biggest challenge was to shed the deep assumptions of reductionist science that had been drilled into my mind for years.

The traditional view of the world — and particularly the living world — holds that reality consists of immovable objects that need to be pushed by an external source to produce movement. This perspective is called causality. For every reaction, there is a cause; without a cause, there is no movement or reaction. Therefore, most scientists still see the world as a system of causal-mechanic interconnection. Causes trigger mechanical motion, like a key winding up a clock.

Poetic ecology, by contrast, proposes that the world is governed, along with causality, by another principle: the principle of self-motion. Objects in the world assemble themselves into larger, more complex forms of their own volition, which originates from within. The particles of matter have a tendency to form bonds and relationships among one another that are mutually transformative. Cosmic dust becomes stars, stars become galaxies, atoms join together into molecules, molecules self-organize into organisms. The principle of self-motion is characterized by the astounding fact that objects start to self-organize increasingly more complex assemblages and then, at some point, develop an active interest in this self-organization and by all means struggle to continue with it. Motion continues not from external causes only but rather from internal purposes: organisms want to live. They strive for a continued existence. We humans are happy to be one of these objects that care for themselves and thus are objects no longer, but subjects: things which have an interest in themselves.

Proponents of a poetic ecology believe that along with causal law, an organic principle exists in the universe that is capable of generating self-animated motion, subjectivity and purposes. They propose that living beings are not an exception to Newtonian principles but rather in themselves reveal the actual character of the world. Some even argue that organic, living reality must be the point of departure for understanding the world and that the world of cause-and-effect is actually a distortion in our perception. The English painter, poet and engraver William Blake, expressed similar sentiments in saying that if we could only see clearly and precisely enough, we would perceive that all things are infinite. We might then be able "To see a World in a Grain of Sand / And a Heaven in a Wild Flower."[1]

Kalevi Kull is a young, restless 63-year-old with a boyish grin and keen sense of irony. When I arrived in the old university city of Tartu, he was just about to travel to his country house in southern Estonia. On the spot he invited me to join him and savor the bucolic Estonian solitude. He sent his young assistant Riste with me to show me the way. While chatting during our journey, she told me of the Estonian passion of jumping naked into the coldest of waterholes at the first sign of spring. And so began my adventure, which led me into the black of nothingness.

My arrival in the newly independent Baltic state some days earlier had been something of a rite of passage, a liminal experience. Rusted Russian planes were parked at the local airport. The road from Tallin, the capital city, to Tartu, the university town, seemed to disappear into a vast expanse of leafless birch trees. The landscape whispered a subliminal, monotonous mantra and enveloped me in its dark vacuum. It smelled of wood fires and snow. Flat wooden houses with faded shingle roofs clustered together to form small villages in scenes out of a 19th-century Russian winter tale. The vast world lay still and bare before me.

I reached Tartu in darkness. There were some bigger buildings in the center and a lot of old wooden houses. Many streets still had their signs written in Cyrillic letters. The yards smelled of rotten potatoes and damp stones. Kull waited for me in front of the university's antique guesthouse, which was made entirely of wood. He led me up the creaking stairs to a spacious attic room. The following morning, he asked me to meet up with him at the former home of Karl-Ernst von Baer, Estonia's most famous scholar, which is now part of the Tartu university campus.

Von Baer played as crucial a role in the holistic life sciences as Darwin did in evolutionary biology. At the time, in the middle of the 19th century, the foundational principles of biology had not yet been decided. Von Baer and Darwin were contemporaries and even corresponded with one another. "If only everyone in the 19th century had not been obsessed with the free market idea!" Kull said ruefully, before leaving me alone in my guesthouse room. "Von Baer could have been another central figure in biology, not just Darwin. If that had happened, our science would have more to do with cooperation and less with competition."

Then I tried to push. I had got out of the Ford Focus and stood in the mud at the front of the car, ready to throw all my weight against it. I waited. Riste needed a maddeningly long time to find herself a position in the driver's seat. I breathed out. When she was finally ready, I groped with my fingers for a dry spot on the cold metal of the hood and pushed with all my force. The engine howled. Nothing moved. The car lodged in the soil as if it was a boulder stuck here since the last ice age. The engine made another howling rev, the tires slogged into the ground, mud splattered against my face, matted my hair and soaked through my trousers. As the Focus slid slightly to the side the hood slipped away from under my hands and I fell.

Riste laughed. I could not hear her but when I had gotten up, I saw her face. Then she climbed out from the other side, opened the car boot and threw her bag over her shoulder. "Come," she said, reaching with her hand towards me. "Kalevi will help us tomorrow." Silence gripped me. My legs were wet and cold. The birches stood motionless. In slender, white rows they emerged from the forest mass, one after the other, a stately procession. Glistening crystals were beginning to form on the windshield. We left the car, our umbilical cord to civilization, behind. Without any sound an enormous great gray owl set off from the tree in front of me, soared along the glade and disappeared into the twilight. We followed the bird into the night.

Towards the end of his life, Karl-Ernst von Baer was something like the Alexander von Humboldt of the North. Along with his biological studies, he undertook vast expeditions throughout the Russian imperial lands in service to the tsars (Estonia was then part of the Russian Empire). Von Baer discovered the mammalian ovum, a major biological revolution at the time, and, like Darwin with his evolutionary theory, also founded a new discipline — developmental biology. This addressed everything evolutionary theory had left out. Von Baer was less interested in how different species came into being than in how the body of every individual evolved in the course of embryonic development. His approach was far less abstract than Darwin's. For von Baer, nature itself formed a harmonious whole, the same interconnected whole as he observed in a single body, which is made up of diverse multitudes of cooperating cells and tissues.

It was a cold morning. Gray and chill. On my way to meet with Kull, I walked under bristly linden trees that stretched the black fans of their branches against a pallid sky. I hurried past buildings on the campus and realized that, in some venerable anatomy lecture theatre nearby, the physician and nature researcher Friedrich Burdach had first used the word "biology" to name the discipline that would become the leading science of the modern epoch.

Kull and I sat down in the former library of the famous professor. Kull tried to explain to me the main difference between the two major approaches towards living beings, the mechanical view and the holistic vision. This latter to me seemed still so incredibly unspoiled in Estonia, and hence Kalevi's position sounded totally natural and absolutely convincing. As weird as a holistic approach may seem in today's world of lab science, 150 years ago the question as to which theory was more adequate to describe organic life was still completely open. The research agenda was basically undefined. Did we need to examine the forces within an organism that led to its development from a fertilized egg? This was the question that Karl-Ernst von Baer sought to answer. Or should we rather try to understand which laws had brought forth such a huge diversity of forms? This was Charles Darwin's focus — and his bold analysis eclipsed von Baer's alternatives for decades.

What are organisms? The standard answer has been that they are clockworks. All aspects of a living being are, without exception, the products of its relentless quest for efficient performance based on how many offspring it can produce in order to propagate and extend more of its genes. Everything else is secondary.

How much did the cultural circumstances and respective intellectual climates of the two biologists influence their ideas? Darwin was a naturalist but still representative of the urbanized civilization of his time. He belonged to the upper middle class, in which, to achieve something and maintain one's status and possessions, struggle was inevitable. England was the most powerful nation in the world, at the very apex of competitive success. Not far from Darwin's suburban villa, sickly yellow pollution belched forth from what was then the largest agglomeration of industry in the world. Great Britain at the time of Darwin was full of desperate poverty. People dwelled miserably in slums and had to compete for the lowest-paid drudgery. One of the significant works of that epoch was "On Population" by the British economist Thomas Robert Malthus. His thesis was clearly inspired by the bleak social situation of the time. Malthus claimed that any population inevitably grew larger, resulting in a brutal, merciless struggle for resources that only the fittest could win. This idea became the cornerstone of Darwin's theory of evolution, which he slowly developed during the five years of his voyage with the *Beagle* and in decades of meticulous fieldwork and reflections. The missing link, which yielded the mechanism of selection, therefore did not stem directly from the observation of nature but from the theory of a social economist, corroborated by his daily experiences of Victorian society. When Darwin hitched up his carriage to ride through London, this is exactly what he observed. He saw the dire social effects of industrial capitalism, and influenced by this unremitting reality, following Malthus, he made a grand conjecture about all of nature.

In contrast, Karl-Ernst von Baer lived in a world of sweeping natural bounty. In Estonia, human beings are dwarfed by the vast silent space of pristine nature. Distances between settlements were huge. Broad stretches of woods and wild meandering rivers were interrupted here and there by immense swamps. Stillness was everywhere, broken only occasionally by the wolverine's snarl, the shrill call of the white-tailed eagle and the roaring of the elks. "Every living being is a thought of nature," he wrote. He was an aristocrat, and therefore from his social position could afford to ponder his ideas at leisure without struggle or competition. Was von Baer less biased in his view of the world? Or biased differently? Was he just projecting a different cultural perspective onto his perceptions of the biosphere?

To this day, Darwin is credited with discovering the principle of how the sheer diversity of organisms came into being. In his time, what seems so obvious to us was quite outrageous. But his ideas ultimately prevailed, convincing a whole reluctant culture, or at least its intelligentsia, that species do evolve gradually from one another and have not been ready-made by a heavenly creator fiddling with a divine drawing board. With the principle of natural selection, Darwin put man back into the web of life. But even if Darwin's ingenious schema restored the human species to its rightful place within the whole of nature, it also imposed a transient cultural worldview — that of the Industrial Revolution in Victorian England — onto our understanding of living beings, in effect distorting our understanding of natural processes. The idea that all diversity arises only through the struggle for survival was clearly a projection of the life circumstances of Victorian England onto the whole of biological reality.

The mechanism that Darwin proposed for his grand idea of a gradual evolution, natural selection, is regarded by some of today's evolutionary researchers like James Shapiro of Chicago University, Eva Jablonka of Tel Aviv University and Gerd Müller of Vienna

University, as too one-sided.[2] It conflated an upper-class perspective of social reality in 19th-century England with a timeless, universal natural law comparable to gravity. Not surprisingly, this force of "nature" was thought to operate like an external power that influenced "objects" such as living organisms. Individual beings are seen as governed by outside forces in the same manner that gravitational force rules a grain of sand. In Darwin's view, living beings become atoms in a vast biological mechanism. By claiming that the species that most capably followed this blind law would produce the most progeny and increase its chances of survival, Darwin seemed to have found a simple recipe for explaining how living forms in all their varieties come into being.

Earlier scholars were already convinced that living bodies were nothing more than sophisticated mechanisms. In early modern times, the whole cosmos seemed to be explicable by physical forces alone. Skilled technicians, eager to prove the truth of that metaphor, went so far as to actually build a mechanical goose that was able to lay an iron egg. The weak spot in this theory, of course, was that one still had to invoke a creator to account for the *design* of all the multitudes of beings in the biological world. However, the whole thing, once set into motion, could automatically play its preprogrammed tune like a player piano. Thanks to Darwin, the origins of this sophisticated clockworks — the selection of the fittest — were also finally revealed as a mechanism, and a most refined one at that. Was God himself, in the end, nothing but an astute breeder of pigeons and rabbits?

At the end of his life, interestingly, Darwin attempted to temper the implications of his own theory, and, to the confusion of contemporary mainstream biologists, he abandoned a rigid interpretation of the law of selection. As an old man who felt profound awe towards nature's beauty, he came to doubt his own famous view that all life forms and their functions could be explained only in terms of their utility for selection. By the time he entered this last phase of his thinking Darwin had long suffered from occasional heavy panic attacks and sudden fears of suffocation that confined him to his room for days. He rarely left his stately Victorian home. He read and he wrote. And he bred exotic plants in his greenhouses, an activity which reminded him of his journey with the research vessel, the *Beagle*. Darwin also studied the behavior of earthworms, among the most humble of earthly creatures. His book, *The Formation of Vegetable Mould through the Action of Worms, with Observations on their Habits*, considered living things as far more than machinery. Instead, he observed that feeling was a pervasive force in all organisms. But by that time, Darwin's reputation had been annexed and transformed by a clique of scientific ideologues who insisted upon seeing natural selection as an all-encompassing view of reality. The old scholar was a shadow of himself, a living monument whose very identity had become synonymous with "the survival of the fittest."

Riste and I had probably been walking for half an hour. The light had seeped away and was now gone. We could not see the path anymore. It seemed as though there were only trees in our way. We stopped and stood on the spot until it dawned upon us that we were lost. We could not even see any direction to go. If we tried to go farther, we would probably go hopelessly astray in the forest. When would the sun rise tomorrow? I asked. Riste did not answer. Strangely I did not feel anxious. Nor was I angry. I felt nervousness in my throat, but it seemed to me that I could still think clearly. It had been thoughtless to leave the car in the forest when dusk was setting in. I told Riste that we better stay where we were. The air smelled of snow and lichen. It tasted of cold and crystalline minerals, as if the trees' breath had given the atmosphere a rare, surreal cleanliness. We were alone.

After a few more minutes, a penetrating chill arrived. I could hardly feel my legs. It was less a sensation of unease or discomfort than a feeling of abject defeat — and fear. It occurred to me that there was nothing I could do. Riste's teeth chattered. We jumped up and down to keep warm but after a while it was too tiring. We huddled together and tried to stay awake.

Kull had told me on our first meeting, "Life is need." How right he was. Right now, it did not seem to be much else. And it was certainly more than an intellectual concept.

"How does the body build itself up during development?" Kull asked, leaning towards me, his old wooden chair screeching. "What is it that makes this body possible? All its coherence? Its meaningful functioning? What force is responsible for the meticulous organization of the cells and organs and all of those structures, so they can finally be subject to Darwinian selection? What holds the machinery together that reads and translates the genetic code? What forces are present in the body that make it able to constantly change and still retain the same identity?"

For Kull, Darwin had told only half of the story, while applying it to the whole. His theory dealt with fully formed bodies. Darwin's selection principle pits mature organisms against each other and, according to Kull, is then erroneously invoked as the sole answer for the mystery of life itself. Even today, Darwin's theory says nothing about how an organism self-creates or governs itself, or by what means it changes. The success of evolutionary theory — and later of genetics and finally the combination of both in the "modern synthetic theory of evolution" or the "new synthesis" of the 1930s and 1940s — made biology the leading science. But even after this triumph, the old, haunting questions remain: What gives the billions of highly reactive substances in each individual cell alone their form? What makes cells separate, unique and highly ordered? What makes all this complex environment behave in a meaningful way? How exactly does the machinery of the living being organize the incredible variety of processes and materials in an unimaginably small space? Which force binds the elements together to form the first primitive cell that struggled to stay intact?

All of these questions speak to a highly enigmatic reality existing in every life form, and so within ourselves. And all these questions are still open to debate. According to Kull, the crucial question that most interested von Baer has still not been answered:

how does a living entity develop a set of activities reflecting its own needs? Kull said: "The theory of how living beings evolve still does not explain *what life is.*"

Until recently, it has not been possible to candidly state these questions within respectable circles of biology. Biological textbooks contain enormous amounts of information about living beings such as their growth, motion, communication and DNA. But all these characteristics do not get to the crucial point. What *is* life? If a young biologist was too interested in such questions during the 1980s and 90s — as I was — he could not have much hopes of a career in science.

Since von Baer's time, a new view of organisms — biosemiotics — has emerged. I was no longer surprised when Kull told me that two of the most influential contributors to modern biosemiotics have Estonian roots — a Baltic-German biologist, the Baron Jakob von Uexküll, and his son, Thure, who founded psychosomatic medicine in Germany in the 1960s. During his active years from the late 19th century into the 1940s, Jakob von Uexküll was an original thinker who did not belong to any of the leading schools of thought in the life sciences. He basically followed von Baer in thinking that it was crucial to research a living being's *activities* and not to misunderstand it as a passive object buffeted by external laws. Dismissed as an eccentric in the standard biology textbooks at the beginning of the 20th century, Uexküll developed a provocative theory of "nature as meaning." His would-be revolution in biology was a one-man endeavor.

Uexküll holds a unique position in the late 19th and early 20th century science wars in which pure materialists fought against traditionalist "nature philosophers" who tried to articulate a principle of "mind" responsible for the special characters of living beings. The old German romantic idealism was making its last stand against the highly successful "pure" science. Uexküll had a position somewhere in the middle, which is one reason that he has remained a marginal figure in biology — and yet at same time the reason for his continued importance.

The Baltic biologist did not view organisms as gene-directed clockworks, as most mainstream evolutionary biologists do today. Many, therefore, dismissed him as a "vitalist" — someone who does not accept that life in the end is nothing but pure matter. Vitalists believed in an extra-mechanical ordering force that did not find a place in pure natural science. This force intervened, God-like, in all life processes. But Uexküll did not believe in a supernatural "Lebenskraft," a special "life force" unique to organisms that distinguished organic beings from "mere" physical matter, as did some of his colleagues such as the German biologist Hans Driesch. The vitalist movement in biology was the last gasp of a dying school of thought before the new orthodoxy took root, insisting that organisms are nothing but tiny machines.

Uexküll, however, took a third direction. According to his ideas, organisms are indeed totally different from inert matter. They have an active interest in themselves and their continued existence. As a result they develop according to purposes that are encoded within their bodies. But for Uexküll, this goal-directedness, this vital self-interest, had nothing to do with old-school metaphysics and otherworldly intervention. He argued something quite different and subtle. Living beings, he claimed, are able to unfold an individual autonomy in dialogue with external influences only because they follow specifications encoded in their body structures (and, as we would say today, in their genes). This liberates them from being slaves to mechanical causality. In Uexküll's eyes, all living beings must be viewed as subjects. Like ourselves.

Uexküll, a member of the Baltic-German aristocracy, lost his estate soon after the Estonian declaration of independence in 1918. After a complicated professional odyssey, he ended up supervising a small laboratory in the northern German port metropolis of Hamburg, becoming an provincial scholar on the fringes of biology. Nonetheless, he deeply influenced many important scholars of the 20th century, including the German philosophers Ernst Cassirer and Martin Heidegger, and the influential French theorist Maurice

Merleau-Ponty, among others. Today, Kull told me, the biological ideas of the Baltic baron force us to rethink the completeness of the "modern synthesis" of biology. "A living being simply cannot be solely understood as a mechanism to maximize efficiency in order to survive in a hostile world," he said.

Then Kull explained the rationale of his theories. I admit that from time to time I had to pause a bit for breath. "A living being is a closed circular process," he told me. "Life means molecules that on their own accord bring forth more of their kind. Life means structures that preserve and repair themselves. Life means flesh that creates flesh, and, first and foremost, does so on its own behalf with no regard for an external world." The living being's total fixation on itself has a noticeable consequence, Kull explained. There is no objective reality that could have an effect on an organism, or which organisms (or humans) could be able to perceive as it is. What reaches the organism is something, a stimulus, an encounter, a push, which gathers meaning only in the light of the organism's needs. Living beings only react to what is important for their existence to continue, regardless of what this really is. Or, perhaps more accurately, what really exists is something which grows together with the ways organisms imagine their relationships with the remainder of the world. They are not predetermined by compulsive physical laws of an external environment but rather by their own inner urges to realize themselves as a feeling body, as a center of experience and concern.

Every being in this manner becomes a sort of creative epicenter of its own world. It imagines its reality by feeling what is meaningful for it to thrive. An organism interprets any influence — whether through the genes or the environment — in the light of its desire to preserve itself. This view is utterly incompatible with traditional Western thinking, which sees a world of distinct causes and effects, and defined agents separated from a background context that somehow differentiates the winners of natural selection from the losers.

51

I felt a little dizzy. Where could I find some solid ground? What are we if not the matter of our bodies? Should every living being itself perhaps be understood as an "unmoved mover," much as the Greeks regarded the divine as a force that gave the cosmos and all things within it their shapes and directions? How could order originate from the *inside* solely by itself, against the convictions of physicists who state that the universe tends toward entropy and disorder, unable to generate its own structures, order and intricacy?

I immediately realized, however, that Kull and the Estonian holistic school of biology had an unsurpassable advantage over the reigning, still more Newtonian, mainstream view in biology: it corresponds to the ubiquitous interconnected cooperation found everywhere in life. Belief in a linear action of genes makes it impossible to grasp these processes. Only the ordered interplay of the myriad of tiny building blocks in each living being's reality can generate the dynamic of living existence. And not only that — besides ignoring certain findings of how the cell obviously works in all its interconnected complexity, the gene-centered view pushes away the experience of how it is to be a living body. I decided to bombard Kull with questions until I had understood everything.

A few days later in the forest, though, without my wanting it to do so, Kalevi's definition became quite personal. Lost in the dank forest darkness, living was what *I wanted*. Here, in that icy gloom between the murky trees, everything inside of me wanted to keep going. To live! As the cold pierced my skin and cramps shot through my legs, I shed layer after layer of the illusions that culture had instilled in me. I was an animal again. I was something that can die.

LIFE AS IMAGINATION

Uexküll's ideas did not immediately spread, and they certainly did not catch fire. This is probably due to the fact that they would have been highly disruptive. Biologists would have had to radically revise their view of organisms and the whole living sphere. Taken seriously, his ideas make clear that biology, while celebrated as a leading science, has still not really defined what it is all about. If subjectivity is the key principle of life, and if the secret of this subjectivity lies enfolded in the body — which cannot then be seen as a delicate kind of robot — then what is the body? What is life?

In 1998 I went to Paris to work with the only biologist who at that time dared to try to answer these questions. Francesco Varela did not simply name some qualities living beings have; he tried to define the process of being alive. I had the luck to be part of that scholar's lab group for nearly a whole year while preparing my doctoral thesis. It was a mind-blowing experience. Shortly before I finished the doctoral thesis I had set out to write with Varela in Paris, he died, way too early. He did not survive the long-term complications of a liver transplant.

Francesco Varela filled Karl-Ernst von Baer's and Jakob von Uexküll's notions of a poetic ecology with conceptual precision — admittedly, without knowing about their work. Varela demonstrated by means of empirical science that living beings must be subjects. He empirically explored the consequences of this new perspective for our view of the world and our image of ourselves. In a heroic effort, Varela's work filled the void — the shying away from defining its proper subject, life — that biology has carried in its center for too long. In a slender paper published in 1997, Varela described his breakthrough observation in one laconic sentence: "Life is a process of creating an identity."[3] A thing lives through the process of maintaining itself continuously as a whole. It strives to regenerate, grow and maintain its boundaries against internal fluctuations and external disturbances. Put in a more radical way: life is not a cascade of causal reactions, but rather its opposite — autonomy.

The young Varela was in many ways a prodigy. Born in Chile, he completed his PhD at the University of Santiago in his mid-20s and was offered a grant to do research at Harvard University. When Pinochet came to power after his brutal coup in 1973, the young cognition researcher left his home country forever. He embarked on a short academic odyssey that took him to, among other places, Bremen, Germany, which he found dreadful. After a while Varela found a permanent post in Paris. Until his death in 2001, he was research director in the neuropsychology department on the vast, gloomy premises of the venerable La Pitié-Salpétrière clinic. Varela occupied a small laboratory in the basement of a 1960s building complex; his desk had a commanding view of a gray concrete retaining wall and the dying remnants of some bushes that a secretary had planted years earlier. Some hundred-odd years before, close to Varela's office in one of the nearby brick buildings with their tall, smoke-blackened chimneys, the famous nerve doctor Jean-Martin Charcot had instructed the later founder of psychoanalysis, Sigmund Freud, on healing women.

Varela's research group, which I was allowed to join, consisted of a couple of highly intelligent young men. Amy, Varela's bright and beautiful American wife, a psychotherapist, called us, with mocking tenderness, the "labo-boys." She captured us quite well. We were continuously debating highbrow philosophical matters while behaving like a pack of young, unruly dogs. In one moment we would mindlessly scarf down our greasy lunch of steak haché frites in the canteen, and in another passionately argue metaphysical issues. All of us PhD students were searching for a link between bodily activity and experience. In the basement of the building, there was a laboratory for human test subjects decorated in a stuffy French style from the 1960s. A coffee machine dispensed the worst coffee imaginable, a proverbial thin lab brew, which we nonetheless needed to get through the afternoon. I was regularly asked to participate as a test subject. With dozens of electrodes neatly arrayed on my head in a cloth hood with a lot of electrolytic jelly beneath, I had to observe the movement of small light points in the dark or recount my perceptions of the slow emergence of picture puzzles.

Varela was trying to gather evidence for the claim that body and mind were not separate from each other, but rather two sides of the same reality. In a series of trials, he was developing the new field of neurophenomenology, whose goal was to do brain research from the inside by comparing electrode scans of biophysical activity with a subject's experienced reality. In Varela's eyes, feelings and thoughts could not be reduced to bodily processes and neuronal impulses. Rather, feelings and thoughts represent the experiential aspect of our body — the side that we see from within, so to speak. This inner perspective may vary and shift over time, yet the experiences nonetheless remain those of a single individual organism and contribute critically to its embodied identity.

For Varela, science's first priority must be to rediscover the egocentric perspective as a constitutive element of life. And then it must accept that the egocentric perspective is not possible without reference to the other, which serves as a crucial counterpart in creating the emergent self. For Varela, living beings are precisely those beings whose actions are primarily concerned with themselves. All other attempts to describe life that overlook organisms' self-awareness, subjectivity and all-consuming fixation with themselves, miss the point. Comparing an organism to a machine is profoundly misleading because it does not acknowledge a being's life struggle to perpetuate its own continued existence. Machines, as humans have designed them, process matter and manufacture things — but do not bring forth themselves. They have no active interests. They never resist being switched off or dismantled. To those who are alive, life is an obsession. Life makes a lump of matter deeply invested in preserving its particular form and its own freedom to act. For Varela, the essence of biological world-making is a living cell, which continuously produces its own components, which are necessary to produce more of themselves. A living cell pursues self-creation. The cell is the physical realization of the principle of subjectivity — its manifestation in time and space, its real presence.

This activity is a feature that cells share with everything that lives. Only the degree of complexity changes. But the principle is tied to everything that is alive. To be alive means to maintain one's own identity as a body, and therefore to have a point of view in the world, an active self-interest in one's own circumstances. Cells spend the lion's share of their biochemical activities unfolding order within themselves, stabilizing it and constructing it anew. It is almost as if the dream of the famous fictitious Baron Münchhausen had become real — in one of his tales he claimed to have pulled himself out from a patch of swamp by his bootstraps. Organisms are bootstrapping all over the place. They assert themselves against the forces of chaos tearing at them. They close themselves up against quantum fluctuations, which can destroy ordered molecules in a merciless bombardment of random breakdowns.

In every moment bodies are engaged in the work of dogged self-construction to counter the ubiquitous downward push of entropy. A single cell repairs hundreds of dissolved DNA bonds every second to prevent its life from coming to a sudden end. Each act of repair is less the preservation of a fixed order than part of an ongoing self-reconstruction. Without their active self-production, cells would crumble into heaps of molecules within seconds. That is in fact what happens in death. Dying means no longer being able to self-create. It means to stop creating. A living being that can no longer bring itself forth as the center of its identity-making activity melts away into a mass of atoms, rejoining the universe of all other atoms.

If we follow Varela, the prototype of all subjectivity, therefore, is a subjectivity of the body, not of the mind. Its innermost character consists of the autonomy of form over matter. The living cell governs the atoms of which it is built. The identity it is maintaining is what binds matter together. This is the reason why, in the view of poetic ecology, another important quality of living beings is inseparable from the already unusual features of subjectivity and autonomy. We can call this quality freedom. Subjectivity is the ability to choose. Choice is coexistent with life. Already a bacterium can select how to behave and how to continue its self-production. The mere fact that it has to become active to obtain food (but also can remain inert) shows that it must select among various possibilities. Even if these choices are not fundamentally independent of the necessities of its soma, there remains a huge contrast to anything not alive, for instance, a grain of sand, which only moves when it is pushed externally. Although a bacterium won't muse over how to pass the evening, as we do, it is not a lifeless mineral. Every cell decides and chooses. It organizes the way it makes connections to neighboring cells, how it puts together its internal structure and outer membrane, and ultimately, as we have seen above, it does so in a way that can become rather independent from genetic instructions.[4] It does so as a way of meeting its needs, producing the actions necessary to remain itself, to keep itself alive and to flourish. Any cell is free precisely because it needs to subsist. We could accurately say: it is free out of necessity.

For Varela, the imperative of self-assertion must be accepted as the common denominator in the physics of life. Self-determination is real. It is an active pursuit. It is the defining moment of living things. For Varela, the machine model of life painfully misses the point: living beings are distinctive for being precisely what machines are not. Machines have no interest in maintaining an identity. In the early 1980s, in collaboration with his Chilean teacher, Humberto Maturana, Varela defined living beings as auto-poetic or self-creating. With this concept, the two biologists had triggered major interest outside biology circles, particularly in psychology and the social sciences, which quickly adopted some of their ideas and through them clearly were inspired to steer away from a causal-mechanic view. But biology is only now absorbing Varela's pathbreaking insights.

MATTER CHANGES, IDENTITY REMAINS

It is startling that modern life sciences have long overlooked the peculiarly active nature of organisms. By ignoring it, however, the pioneers of biochemistry could proceed more quickly. They viewed cells as gigantic chemical reaction equations that they could precisely represent and manipulate. Step by step, scientists reconstructed the reaction chains for all metabolic pathways in living organisms. In eighth-grade biology, most of us learned about some of these reaction networks and undoubtedly found them rather dull — and quite hard to memorize.

The physiologists mapping the reaction pathways inside cells were of course following a totally deterministic methodology. The irony is that they could just as easily have observed the ways in which cells are *not* causal chains. The proof for it was a central part of their work and still is part of basic biochemistry high school teaching. The principle I am talking about is neatly delineated on the typical, colorful metabolic pathways posters displayed on so many lab doors — often as a marketing giveaway from corporations selling chemicals used in physiology experiments. These posters portray the jungle of biochemical reactions in a cell. In some respect their meshworks resemble highly complex, integrated electronic circuits. Maybe you remember one of those from school days?

In the center of the cell metabolism is the citrate or Krebs cycle. This metabolic hub provides for cellular respiration and, therefore, is inseperable from a cell's life-creation process. The Krebs cycle shows how an organism takes up sugar to feed on the energy contained in it and how it emits carbon dioxide when that energy is used up. If you look carefully at the cycle, you discover a peculiarity that is never taught in school and which scientists conditioned by the prevailing paradigm have largely ignored.

The "C" — the carbon atom entering the cycle and hence the cell body in the chemical form of Coenzyme-A — is not the same atom as that which is exhaled as carbon dioxide. Rather, the carbon in the sugar becomes part of the cycle and hence part of the organism's body, ready to be transformed into anything a cell can produce, from a lipid membrane to a moving element in a sperm's wiggling tail. The "C" in the carbon dioxide, on the other hand, is not simply used-up and discarded fuel. It is an integral part of the body which, upon exhalation, is absorbed into the surrounding world.

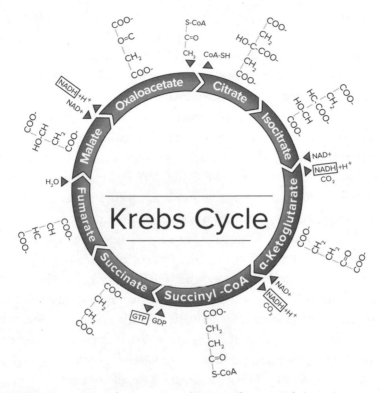

We must not underestimate the significance of this observation because the particular reactions of the Krebs cycle help us uncover the character of life itself. First, we can see that life is not a linear chain but rather a loop reproducing itself. The building blocks that a cell brings forth — the proteins in its interior, the fats and phosphates of its membranes, the sugar chains on its surface — are created by its metabolism, but at the same time they comprise its metabolism. The building blocks produced contain the very means needed to create more of themselves. We can see that an organism, through its metabolism, creates a partly independent, closed-loop system, as depicted in the cyclic Krebs reaction loop.

Therefore metabolism does not simply mean processing something external to one's self, using it up and then shedding what's left in an altered form. This prevailing view is an objectivist distortion of what the cell is really doing. Metabolism is not anything like an oil change in a motor, nor can we understand its function by comparing it to a gasoline combustion engine and the release of exhausted fuel. Metabolism, a term borrowed from Greek, literally means "overthrowing." It is not the process of taking up energy-rich nutrients and then discarding what remains. Metabolism means tearing down one's own physical substance and building it anew with what comes from the world. Metabolism means sharing one's own matter, and hence a part of one's own identity, with the world. When I eat, my use of this "fuel" is quite different from a motor's use of fuel. An engine burns fuel to produce energy and move its pistons, a process that transforms the fuel into CO_2, which is then released as exhaust.

But the food that a living being takes up becomes a part of itself. And as a consequence, what is shed after the energy has been consumed is not a waste product but an organism's actual substance, part of its bodily identity. In ordinary respiration, cells release piece after piece of themselves. With every breath, we all surrender fragments of our bodies, constituents of ourselves, into the surrounding air. At the same time, we all continuously incorporate and transform many elements of the earth into ourselves. We become what has been the environment. What I am now, moments ago was corn on the field. What I have just been is now already carbon dioxide in the inner space of a leaf's loose green tissue, and then it will be part of a blade of grass on the meadow. Any creature that I eat is literally now what I will be then. The "it" will become "me."

But what this "me" is becomes rather complicated. I cannot command the stuff of my body. What makes up my physical identity is "not me" but rather the matter that other beings have consisted of. Through my food I simply transform the stuff that the whole world is made of into me. As poet and novelist James Hamilton-Paterson so precisely observes, we are literally made up of the stuff that was once in the center of the Big Bang. A deep part of my self is *not* myself. My body does not consist of my own specific particles because the matter that makes up my body is constantly changing. The changes in my body physically connect me to all other particles in the universe. An organism's constant intercourse with the universe builds it up through that which it is not, transforming it in the process. I, like every being, can build an identity only because I do not truly own myself.

Over the last two centuries, biologists had little interest in such ideas because they were so intent on developing highly precise methodologies in order to produce predictable results that could be considered rigorously scientific — just as physics was doing with such great success. Biologists became happy Newtonians, and most remain so to this day. And indeed, by slavishly adhering to the aims of a naïve objectivity, exactitude and reproducibility, biologists have made huge achievements. Ironically enough, precisely by being good Newtonians, researchers have accidentally discovered how living beings cannot be described by a deterministic physics. Just think again of the Krebs cycle: in order to trace its functioning scientists needed to do hands-on reductionist lab research. They worked with the assumption that cellular metabolism must be understood as a causal chain. But their research yielded results that, upon further interpretation, point beyond the mere causal functioning of cells. Many biologists are realizing that organisms cannot be understood as physical machines, simply because that would violate a principal law of machine design, the principle of material self-identity of any machine, like, for instance, a motor. Machines do not metabolize; they do not morph into the matter that surrounds them as living beings do. Machines substantially remain the same day after day, and even after centuries you could take a sample of their matter and find that the atoms have not changed. To see the crucial difference between organisms and machines, it was necessary for biologists to spend generations believing the creative misunderstanding that organisms *are* machines. Only by applying Newtonian thinking so diligently could biologists have discovered that living beings in fact transcend the causal mechanics of Newtonian thought.

Nobody expected this to happen. The founders of modern-day biology could not have predicted what would finally come into focus after decades of developing the research fields of physiology, biochemistry and genetics with the mindset of physical science. They could not have predicted the ways in which biology, in the end, *did* become like physics. Paradoxically, real-world biological phenomena do not resemble the lawful, mechanistic, linear simplicity of Newtonian physics and its eternal laws of gravity and acceleration. They instead manifest the same problems we find in the weird and subjective world of quantum physics. It turns out that every subject — every organism — is inextricably connected with the totality of the world, and that subjectivity cannot be excluded from the understanding of life. Subjectivity is the door to a deeper knowledge of it.

In this recognition, biology finally becomes truly modern. The downside, for the moment, is that biologists are now confronting a deep paradox similar to the one that physics has struggled with for the past hundred years. I want to emphasize, however, that the correspondence between biology and physics is not like those esoteric explanation models that glibly use quantum-talk to explain, for instance, consciousness. Following Varela's research and his determination to stick to what we can observe, I think that a physical basis empirically (and mathematically) for linking modern physics and a future biology is not yet possible. The existing theoretical proposals are mostly too simple and too reductionist.

Where the two fields do converge, however, which we can confirm with a quiet confidence and without the distortions of easy metaphors, is in the role of the organism-as-subject in relation to the environment. Biochemistry, if interpreted in a new light, can show the reality of subjectivity and an individual's point of view, and the need to abandon objectivist dogmas in the contemporary sciences. We have to refrain from neatly separating agents and their environment, and experience and reality; as we have seen earlier, the two are tightly integrated. This is as true for modern physics as it is for biology, although on totally different grounds (at least speaking for the moment and based on secure phenomenological knowledge). In quantum mechanics, matter is neither particle nor wave, but in some strange way both at the same time. In the emerging poetic biology, a living being is neither matter nor form, yet it is a subject that can govern the connection between both. Life is a process that can really turn bread into flesh, bio-chemically, and without any dollops of mysticism added. Matter passes through an organism's body and surrenders to a fixed identity only for short fragments of time. Matter metamorphoses: first it is earth, then you, then me and then the earth again. From this perspective, infinity is at reach in every moment. It flows through us. Every living being is a knot that ties the whole. We are here and everywhere at the same moment.

CENTERS OF VALUE

The fact that any living being's metabolism is structurally open to the world requires the organism to be relentlessly creative. This openness bestows the organism with a fierce drive to invent itself anew in every moment — to produce an "imaginary surplus," as Varela once put it.[5] The need to imagine is intimately connected with its opposite: the constant threat of death. The creativity of embodied existence requires death as its constant companion. An organism's capacity to live or die is less a question of its material composition than a question of the degree to which it can continue to sustain its motivational spark to self-create. A dead organism basically consists of the same matter as a living one, but this matter has come to rest and does not pursue any internal goals. Dead organisms are quickly incorporated as food into the bodies of other organisms that are still able to maintain their capacities of self-creation. Decay is to be integrated into the metabolism of others to generate new life.

Paradoxically, it is exactly the constant threat of death that becomes the impetus for existence. Life needs to be able to fail in order to prevail. Health for an organism does not mean to be undisturbed but rather to be energetic enough to reimagine itself in the next instant. The process of life does not run as smoothly as a well-oiled machine. The process of life is a continuous process of creation struggling against the pull of entropy that governs matter. Atoms on the one hand tend to take on the lowest level of energy. If left alone, they decompose into small, single parts without any activity. But, on the other hand, matter seems to be dreaming of becoming something complex and transforming itself. If we wait long enough, matter can't help but build more complex structures, new connections and different relations. We will delve deeper into this tendency in the next chapter. Here let me only observe that living cells actively harness this potential in matter to self-construct more complicated structures. The constant regeneration of cell organization, the ongoing production of building materials and the repair of DNA, has no other purpose than to keep tired matter going and to eternally regenerate its organizational identity. We cannot understand this creativity if we do not see that it is directly related to the possibility of non-being. Living imagination, in a dialectical manner, is the triumph of the longing for a fuller being over the threat of decay. Every organism, in every instant, is thus a direct embodiment of the triumph over death. At the same time no living being is thinkable without death. Without the constant threat of imminent failure, there would be no desire for cohesion and unity — and therefore, no subject whose identity is only formed by asserting itself against the menace of breakdown.

Only an organism's structural incompleteness makes the formation of a self possible. Only death waiting around the corner provokes a self to close itself up against dying and develop a perspective on the world and a determination to live. The tendency of stuff to complexify itself is not enough to generate life. It also needs the opposing drag of matter towards entropy, to stasis in inertia. Complexification and inertia are both necessary qualities of reality. Only together can they create the tension from which the living can derive its imaginary dimension. The possibility of failing, of dying, is what binds every form of life to an interiority, to a perspective from within. This subjectivity is simply the point of view that focuses and interprets the encounters an organism is faced with. A point of view is essentially about avoiding the detrimental and seeking the advantageous. But introducing a point of view into the interplay of matter changes everything. A life that only a moment earlier had been the meandering of indifferent molecules through different chemical phases can, with the evolution of an interior self and point of view, awaken to value and meaning.

With this, any objectivity is entirely ruled out. To preserve oneself as form over matter means that the living system has a necessary stake in its continued existence. At the same time it means that the organism does not perceive the world "as it is," but only as it matters to its own ongoing existence. Whoever has an interest in continued and even enhanced existence understands the world not "as it objectively is" but only as it reflects his own needs. The world, then, is *imagined* rather than observed. And this imagination is in the service of a living being's own needs to flourish and thrive, which is all it needs to know. A bacterial cell follows a trace of sugar not because it recognizes its components as carbohydrates and therefore, objectively concludes that it must be food, but simply because it has a positive value for it. It does not follow "sugar," it follows "good." We experience this powerful magic ourselves every day, the immensely positive value that comes from satisfying hunger. On the surface, the bacterial cell's behavior appears to be quite simple and automatic, which tempts an observer to compare a cell to a robot. But its simple and efficient behavior is possible only because an organism is capable of goal-oriented action and does not act like a robot. Robots would need an accurate map of the space they are acting in. It is this requirement that in truth turns out to be complicated. For the organism it is much simpler. It can satisfy its needs without a map because it acts out of self-concern. This mode of perception allows for orientation in any environment. There is no need for a map to direct behavior. A living being only needs desire and inner experience.

Feeling is the most accurate way to relate to reality. It might even be the only way (if we regard objectivity as a collective fiction that disguises feeling).

All these insights have the capacity to deeply transform the life sciences. They convert biology into a science of intense experience. If the behavior of a living being can represent the most original forms of subjectivity, so, too, value is coexistent with the physical organism. A living being is value that has become solidified as a body. The experience of value becomes coextensive with life itself as every living thing constantly creates values by having needs that must be fulfilled. Experiencing value from the standpoint of a self becomes the prerequisite to the physical construction of an embodied self. Value is intimately linked to physical reality. For a being in need, the world is not a neutral place but suffused with value: everything assumes a specific value for the organism's future existence. And only an organism in need can exist at all. This is true for us as well. And for the snake fly. And for the budding mustard germ. And for the *E. coli* cells in our guts. All these can exist as living systems only because they bring forth value and inwardness in the physics of life.

For centuries most biologists considered any living being apart from humans as free of value; they were seen as neutral and no different from a clockwork or a tiny gearbox. For most scientists, value still is a term that can only be used in the context of human morality, which sets us apart from the rest of life and hence is not seen as a topic for biology. Similarly, to most philosophers, the world-making of other living beings and the structures of ecosystems are ethically mute. They may exist physically but offer no valuable ethical meanings. For centuries, natural facts have been regarded as mindless and causal-mechanic — without moral implications. Anyone who disagreed was accused of the naturalistic fallacy. But what if organisms as living artifacts of nature *already* exhibit behaviors that speak of value and good and bad experiences? Life, in the view of poetic ecology, is never a neutral place, not even for a single cell. The two worlds of matter and feeling have long been seen as completely separate realms. But now, under the eyes of a poetic ecology, they come together again.

In the emerging new picture we can also reevaluate the image of nature as a pitiless struggle for survival — "red in tooth and claw," as Tennyson famously wrote. The self-interest of living beings and the determination with which they strive for an advantage must no longer be seen as a blind result of the gruesome, imperious logic of nature. We are invited, instead, to consider how the egocentric perspectives of living beings give rise to creative autonomy, quite the opposite of blind forces and mechanical causality. Whoever acts with self-interest has a self to defend. This self is unquestionably not "ego" in the sense that psychologists speak of, but it is certainly a center around which an organism organizes the world and its experiences. The interior self develops its own consistent worldview and identity. Only by experiencing their own existential needs do living beings give rise to the value categories of "good" and "bad." A canyon is indifferent to whether it is washed out and eroded by the flows of water over the course of millions of years — or whether it is filled in by the blowing sands of a dry desert. A fish, however, whose familiar stretch of river is drying up, will struggle for its survival, twitching and whipping its silvery body as long as there is life in it.

If we acknowledge that life confirms the experience of value as a physical force, then everything changes. A factor that we have reserved for ourselves with near exclusivity, can be recognized as a literal physical force affecting the reality of life — the force of feeling. Feeling is the condition by which subjective meaning is experienced. Feeling is an existential message of meaning from the self, from an inward perspective. Feeling is the first-person view of existential experience, a measurement of the outlook of an organism and of the degree to which it is coping with its role in the world, through self-creation, self-maintenance and survival.

Therefore, feelings are always a yardstick for good or bad. At the same time, they are not arbitrary or "just subjective." Feelings express the real situation of a particular living being. In this sense they have an objective character. Feelings *assess* an objective situation from the inside of one of its participants. This is why they are always right from an existential standpoint, and why it is so important not to override them. They are the sensing system of our contact with reality. When we remember situations in which we registered distinct gut feelings and yet neglected to heed them, we know that our embodied knowledge has been correct in the end.

It might be even simpler to say that subjective meaning *is* feeling. *Life* is feeling. A cell is feeling expressed in a bodily shape. Feeling is the immediate impression of being alive. It is being alive from the inside. It is the magical transformation, the miraculous experience of catalyzing an identity from physical matter. Instead of seeing rational thinking as the only certainty available to us, as philosopher René Descartes famously declared when he said, "*Cogito ego sum,*" (I think, therefore I am), feeling includes all living beings. It transfers the invisible red line separating our world of sentience from the realm of physical being farther down towards the sphere of the dead, which has no interest in itself. But we should also be careful to remember that the particles of matter are mysteriously social and prefer to explore new connections, even if on the outside they always seem to be merely physical.

The world that poetic ecology describes is vibrant with feeling. From the lowest single-cell organisms onwards, all life is *animated.* We should recognize this fact, directly and rigorously, and give new force to this term. Being animated means being governed by feeling and the desire to grow; animation is the biological power of self-construction and continuity. This is the true nature of feeling in nature. This is the feeling we share with all creatures — the genuine experience of good and bad, the deep waves of misery and joy. Our familiar human self-consciousness is a very late arrival in the long cavalcade of the physics of life. Indeed, human consciousness so often misses the point. In this respect, Aristotle had an interesting suggestion. Observing the ordering power of life and its living form, he called it the soul or psyche. Seen from this Aristotelian angle, soul or psyche is the animating principle of every sensitive, striving body. It is inwardness that gives matter a form and identity over time. The capacity to cradle a soul in this respect cannot remain the exclusive privilege of humans. It is, after all, an empirical biological principle of the living world. Soul originates in the same instant as life. It is its innermost reality. And therefore, it is also its most pervasive expression.

I blinked as the light beam hit my face. At first, the blaze was still part of a confused dream. But the flashlight pierced deeper into my consciousness and brought me slowly back to the present moment, against my will. Kull was there. He said something that I did not immediately understand.

I stared at the scene around me. The shapes of the trees became distinct against the sudden brightness and slowly assumed their independent existence again. Reality was given back in tiny fragments, and we made stumbling steps towards them. Riste's teeth still chattered. Kull did not look concerned. He smiled kindly. The road had been very near all along; it extended behind the third row of trees, as if civilization had never lost sight of us. The professor's car waited, with engine running, several yards away. Hot air hissed from the heating vents, which were turned to maximum volume. When we had not arrived, Kull had decided to drive in our direction. Then he saw our car.

It took only five minutes to drive to the farmstead, which was faintly illuminated by the headlights as we veered into the entrance. There were several low buildings, gray wooden houses with low gables embedded in the soil as if they had been sitting there since primeval times. Kull had recovered the farm after the Russians had left Estonia. The area had belonged to his family for centuries. The houses were made of weathered tree trunks. The gray surface of the wood was covered with a light fluff, making them feel like the skin of a living being. Riste and I burrowed under mountains of slightly damp blankets. Kull started a roaring fire in the fireplace. Our stupor started to fade. We slowly started feeling better. He even conjured up some coffee from a dented metal tin.

In the morning a fleeting sunlight poured a shy glow over the faded wool blankets of my bed. In the afternoon, Kull heated up the sauna. He stacked an enormous pile of birch logs in the low, sturdy hut. Soon, the flames licked the jet-black trunks of the building and thick smoke gushed through the air holes under the roof. I walked to the small lake a few steps away. Its surface was still frozen. It reflected the pale afternoon light, generating a layer of diffuse color between the barren trees. An old shaky footbridge led out a few yards over the water. Ice had formed compact vertical needles, which stood erect like pencils close to one another in the dark water. I carefully pushed my hand forward among the icy mass, and its chill penetrated my wrist. The frozen needles around my skin floated back into place as if my fingers were part of them. I quickly drew back my hand. I'd had enough cold for the time being.

I gazed over the lake. A large beaver lodge arose from the shore, stiff and empty as an abandoned castle. Fallen birches marked the area where the animals had started to engineer a new dam across one of the small creeks feeding the lake. Thin trunks still held themselves upright even though eager teeth had narrowed the tree stalks into hourglasses. I could see from the hue of the ice further in the lake that the sun was about to set. The glow engulfing me resembled strawberry milk, becoming more translucent as darkness descended. The frozen surface of the lake was transparent and without color; it soaked up the sky's shade and surrounded me with a crystalline light that felt like the inside of a seashell.

I touched the crystal needles below the footbridge again. It was as if I had raised a hand into the sky, an infinitesimal movement in boundless space.

CHAPTER 3:

The Physics of Creation

*Lots of simple agents having simple properties may be
brought together, even in a haphazard way, to give rise to
what appears to an observer as a purposeful and integrated
whole, without the need for central supervision.*
— Francisco Varela

I remember a long time ago when on a similarly cold day I was re-turning from a walk. I had been rambling on a sandy road along the fringe of Kisdorfer Wohld, a forest in the north of Hamburg. Over the puddles in the tracks left by tractor wheels delicate layers of ice had just started to build. Subtle feathers of frozen crystals reached out in slender curves and plated the surface with a slowly hardening veil. I stopped short, mesmerized. The icy hoar made me think of a picture I had seen above my head only some moments earlier: the pale blue sky being consumed by growing cirrus clouds. These clouds, as I knew, were made from ice, even though they existed at an incomparably larger scale. In front of my eyes the sky rose a second time, in the puddles. Was I gazing down or up?

At the edge of one of the small pools I discovered a grayish-blue feather from a common wood pigeon. It seemed related to the icy layers in the sky and on the water in the sand tracks. The branches and barbules that form its surface are themselves built like minuscule wings, also looked as if they were crystallized into shape.

These and other patterns have increasingly convinced me that life and its non-living, inorganic environment are closely related. To me, both dimensions of reality, while obviously different, seem less and less separated by an insurmountable abyss. And indeed science has found a host of examples that bridge both spheres. Researchers today can not only comprehend which of the simple patterns already active in nonliving matter but also create structures in organisms. They have even understood how such a non-living, anorganic ordering process is able to shape matter of its own accord. The principle central to life, the fact that the greater whole decides the fate of the matter involved in its construction, is not only active in living beings. It already shows itself through complicated arrangements of particles. If the mixture of matter is sufficiently complex then it tends to preserve this complexity through its own tendencies to increase the meshwork of inner relationships. We cannot any longer deny this observation — subjectivity acts as a physical force. It can even be proven in the lab. Maturana's and Varela's autopoiesis idea has found overwhelming support from complex anorganic systems. In this short chapter I want to examine more closely the physical, biochemical and genetic foundations of a poetic ecology.

"WE HAVE ALWAYS UNDERESTIMATED CELLS"

For a long time scientists believed that biological order could only be achieved through external forces acting on the chemical soup of the cell, for instance by instructions from the "genetic blueprint." This was a secular successor to the old idea that a heavenly Creator intervened through subtle mechanical adjustments. Living matter was thought to be passive and waiting to receive interventions from a causal force, as in Newtonian physics. In the last 30 years, however, it has become more and more evident that many biological structures come about without any outside plan or control. They self-organize. Chemical reactions automatically settle into energetically stable states. In the test tube, researchers observe self-creating patterns of waves or circles that cease only if one of the reagents involved is used up. Many regularities in nature arrange themselves. The most famous example is probably the singular order of the ice crystal's filigree, which arises without any instruction. In organisms, too, many features are self-organising: the zebra's stripe pattern, the leopard's spots, the veins in a leaf, the loop design of our fingertips. None are genetically fixed or specified, but rather emerge on their own according to a certain set of initial conditions. They unfold in the same way that cells grow during their development, following certain general rules of spatial and temporal arrangement. Some delightfully complex structures can arise and develop, but only if a sufficiently large number of single building blocks is involved.

The American biologist and system researcher Stuart Kauffman has explored the rules leading to this amazing self-organizing complexity. He found a seamless transition from self-arranging chemical complexity to the physiological self-construction of living organisms. The most important result of Kauffman's explorations is the claim that the ways in which complexity organizes itself may also explain the origins of individual autonomy. Autonomy, the defining criterion of organisms, seems to emulate the inherent tendency of matter to bring forward self-sufficiency through the creation of individuals. The more highly evolved a system, the stronger its capacity to shield itself from its environment and to develop self-referential behavior focused on the maintenance of the system itself. In this fashion, self-organizing systems can insulate themselves from the volatility of the exterior world and arrange their more complex structures around the intricacies of their own interior relationships. We, therefore, could say that complex structures develop a certain self-centeredness. We could also say that mere matter already shows a first tendency to manifest its own subjectivity, broadly construed. Figuratively speaking, matter opens the door to let selfhood in.

The first decisive step towards subjectivity is what scientists call "autocatalysis." We could translate this term as "self-enhancement" or "self-acceleration." Simulations have shown that the more complex a system is, the less independent its parts are from one another. The parts necessarily start to interact. They catalyze or block one another's construction, so that the chemical soup is no longer a random distribution of all the particles it contains. Instead, defined structures emerge, because the substances which come into contact with each other react in different ways. The mix is not random anymore, but has certain defined qualities which depend on the ingredients present in the mixture and which through their interactions bring forth new qualities of the whole.

Here the power of interrelations becomes all important. In building up structures, some molecules in the process simplify the production of others. These molecules are called catalyzers. Through their spatial and electrical qualities and the ensuing chemical conditions, the interactions of these molecules end up reducing the amount of energy for a certain reaction that would otherwise be needed. The system comes to rely on the new, more complex scheme of reactions, which then establishes itself as a kind of primary, default configuration in the soup. Catalyzation is one of the founding principles of organic existence. Biochemistry is cooperative. If one biochemical molecule is transformed into another, this process almost never happens in isolation but in the presence of, and with the aid of, other species of molecules. Today we know thousands of these catalyzers. All the enzymes in our body, for instance, work this way. They lower the reaction thresholds in order to break down alcohol, for example.

Although scientists know a lot of specific biochemical catalyzers, it could be that we still do not fully understand their crucial roles. Instead of being a helpful biochemical mechanism for specific reactions in organic exchange processes, mutual catalysis might rather be the way *any* relation between the material compounds within an organism is brought about. Kauffman assumes that many more chemical reactions are in a cell than we imagine, catalyzing other reactions or making them possible in the first place. Kauffman claims this is the only way that we can explain the complicated and extraordinarily entangled metabolic pathways in a cell. Not just some biomolecules, but thousands or hundreds of thousands of them — in fact, basically all of them — act as catalyzers.

This idea follows a simple logic which Kauffman has derived by calculation and modeling, however, not by direct observation of the reactions within a living cell. The more classes of molecules you mix together, Kauffman assumes, the higher the chance that one of them will catalyze other possible reactions in this stew of organic possibilities. The magical tipping point is reached when so many molecules are swarming inside the organic broth that every variant catalyzes a reaction leading to the production of another variation of molecules, which in turn catalyzes yet another reaction. At this point a phase shift happens and the molecular network migrates to a higher level of complexity that stabilizes itself in regular reaction patterns. Through extreme complexity, the building blocks in the biochemical soup may become able to instigate their own coming into being, or autopoiesis. At this point there are at least as many reactions bringing forth others as there are classes of molecules. Over time the existence of the entire metabolic unity is secured, and an astonishing order arises in the tiny volume of intracellular space.

For too long scientists have been fixated on what happens chemically within any of the distinct transforming reactions of the metabolic pathways. They elucidate the biochemical balance of the entire metabolism made up of many chemical reactions, but they do not see the emerging whole, which is more than the sum of the reactions because it is constantly assuring its own existence. From the physiological viewpoint, the cellular household was separated into different and isolated causal chains. For this reason, very few researchers have reflected on systemic questions: In what ways are the different reaction chains related to each other and to the system, and what feedback loops can arise among them?

I very well remember an afternoon during my undergraduate studies in Freiburg when I attended a seminar called Philosophy of Biology, which a sympathetic professor with a research focus in genetics was offering. Once a week, we talked about organisms and we read Italian poetry. I especially remember Salvatore Quasimodo and his famous lines, "Everyone stands alone at the heart of the world, / pierced by a ray of sunlight, / and suddenly it's evening." I asked the professor why every molecule in a cell does not react with others all at once. After all, they are all energetically highly charged biomolecules, far from equilibrium, and therefore predisposed to pass on their energy. I felt stupid asking a question whose answer had to be quite obvious; otherwise we would have learned a theory for it in our studies. But the professor had no response.

"Compartmentalization," he finally said after some reflection. By this he meant that everything in a cell is well isolated in miniscule containers and distributed in tiny production chains, as in a factory — an idea I discussed in the last chapter. I accepted the professor's answer, but I was only half-satisfied. Or rather, I was disappointed by biology. How boring organisms were in the end! At the time I did not grasp that the professor's answer only shifted the question to another, equally murky level, namely, what organizes the order of the production lines and tiny compartments? I had been asking the right question after all. In truth, biomolecules do not behave at all like component assemblies on a factory conveyor belt, which are processed step by step. They are rather like manic visitors at an oriental bazaar who buy something at one booth, then have a chat over at the next stall, then greet someone somewhere else. "Catalysis is everywhere," as Kauffman puts it.[1]

This view is supported by American cell biologist Bruce Alberts. "We have always underestimated cells," he says. "Instead of single proteins colliding by chance, every important biological process is achieved by swarms of ten or more protein molecules which are interacting with several others of such collectives."[2] A cell, therefore, is not a water drop in which some proteins swim, but rather a gel whose billions of tightly entangled constituents constantly have numerous contacts with one another. If we follow complexity researchers like Kauffman and Alberts, therefore, we can no longer imagine the cell with its billions of molecules and hundreds of thousands of metabolic pathways as a miniature factory filled with countless test tubes hosting utterly separate reactions. To the contrary — the cell is the opposite of an industrial process. Reactions cannot possibly remain isolated and separate, if only because that would interrupt the constant play of interrelations. What brings forth the complex autonomy of a living being is the restless, robust intermingling of a messy swarm of molecules. The cell's body is a dense matrix of tens of thousands of chemical substances that become so tightly enmeshed that the resulting coherence becomes stable, able to tolerate external assaults and internal fluctuations. No single substance in the cell is guided through a series of different containers; the cell itself passes through a sequence of conformations through which the whole continuously transforms itself.

Many empirical observations support this idea. For instance, living cells polarize light. If light beams are sent through cells, the rays come to vibrate in a uniform direction — evidence that this "biological lens" has a high order of complexity. Chemicals put into solutions in which particles are randomly distributed do not have this effect. This has led some researchers to view cells as liquid crystals. They are fluid, but they have a crystalline order. The botanist and philosopher Anton Markoš of Prague University even hypothesizes that the quasi-crystalline structure of the cell is continued throughout the whole body of an organism. Its distinct units are linked through a narrow meshwork of pores, corridors, hormone channels and electrical contacts. At the same time they are flooded by the liquid that embraces all tissues, the so-called extracellular matrix. Blood, lymph and cerebrospinal fluid wreathe every cell and leash everything together in one giant continuum of a macroscopic liquid crystal. In this densely interwoven state, the organism corresponds less to an industrial installation with its separated production lines than to an electrical or magnetic field. A stimulus at one point in the system can be perceived at any other point instantaneously. This could be the reason for the fact that if a muscle stirs, an astronomical number of molecules of ATP, the energy carrier of the cell — some 1,020 of them — is dispatched and absorbed in a single instant through the muscle. All the cells receive and break down ATP instantly as if the muscle was one huge, integrated unit.

If these hypotheses can be more fully confirmed, then they call into question the framing assumptions of our empirical observations. We would be compelled to observe all constituent elements of a cell simultaneously to understand its actual functioning. Such a reorientation would shift our attention from cause and effect to the active, dynamic balance that the participants of a huge, coherent whole somehow achieve. This also means that we should no longer understand genes as centralized command centers sending orders to outposts in the cellular landscape of an organism. The entangled crystal of a cell includes the DNA, which in turn must be understood as crystalline in its own right, and as structurally related, not causally chained up, to the body of the cell. DNA already has some qualities of a crystal. In higher cells it consists of a few giant molecules with fixed spatial conformations — the chromosomes. Their substance is also part of the highly structured autocatalytic reaction that characterizes the tiny lump of gel inside its membranes that is a cell. A gene is not a more or less isolated lever in a central control unit, but rather one important element in a much larger, more encompassing constellation of organic activity both within an organism and in its immediate environment.

AUTONOMY IN THE UTERUS

All these observations force us to conclude that the behavior of matter cannot be categorically separated from what happens within a living being. But that's not because living beings function as cause-and-effect machines, as matter supposedly does, but because matter itself has an inherent drive to plenitude. The behavior of molecules in the interior of a cell evinces the beginning of that autonomy — a trait that Varela regarded as the central trait of organisms.

Whether this view can explain many enigmatic features of different organisms remains to be seen, but scientists are adducing more and more evidence to support this hypothesis. Developmental biologists can now demonstrate that during early development, an organism behaves more as a subject than as the product of genetic machinery alone. Many fields in developmental biology today directly support Varela's ideas. For many years, however, the principles of embryonic development were inscrutable to biologists. Which forces guide a shapeless, protein-rich liquid inside an egg to become a complex, highly differentiated organism? What causes one drip of protein pulp to form a leg and another a wing? The most logical answer is that a developing organism has to be understood as a complex whole that organizes itself and forms an autonomous network of internal reactions that mutually maintain one another.

The genes play a central role within this complex whole. But this role is different from what researchers had thought. "The miracle of complex life is stranger, but ironically also much simpler than anyone could imagine," the prominent American embryologist Sean Carroll claims.[3] Contrary to what most scientists have believed, genes do not contain the complete building instructions for constructing a living being. The hereditary material is comprised of many different elements and unconnected modules that, by themselves, could never give rise to a cell. For a cell's form to materialize, a second, feature of the genes is important — gene switches. DNA does not only consist of code sequences for proteins, it also consists of many sequences of DNA code that can activate or shut down other parts of the genome. DNA, therefore, is not only an archive of information but also a highly complicated computer executing calculations through massively distributed networks that function according to completely different principles than any human-made artificial cognitive system. Which genes in a cell are active and which generate proteins in this setting is less dependent on an unambiguous "genetic order" than on the outcome of highly complex and nonlinear interconnections.

We here encounter a situation similar to the interior of the cell, the site of a vast chaos of chemical reactions self-structuring itself. Again, the genetic switches must be understood as a network establishing its own structure through the vastness of possible interactions. The pattern emerging from a huge variety of interconnections feeds back on these structures themselves and thereby influences their arrangement. Hence, it is impossible in principle for preexisting genetic information to dictate outcomes if this information only arises *after* a certain complexity threshold is crossed. Rather, the state of the network itself decides which shape the emerging organism shall take. But this can only happen on the fly in an active, developing network whose outcome is theoretically impossible to compute in advance. Only a real somatic network that is actually traveling through space and time can yield this outcome. Once switched on, emerging tissues follow the paths that make sense to them. Here again, the tissue's trajectory corresponds less to simple chemical reactions than to more complex behavior. Cell groups differentiate, for instance, into liver tissue, if they are in sync with the right stimuli from their environment. Like a songbird building his nest only when longer daylight hours and warm sunbeams have raised the hormone level in his blood, cells respond to the messages of DNA according to their individual situation.

"Weak linkage" is the term the Harvard biologists Marc Kirschner and John Gerhart have coined for this stubborn indeterminacy of developing bodies[4]. With this expression the two researchers have stepped out of the causal-mechanic paradigm that has long prevailed in biology. Causality always presupposes strong linkage. One thing mechanically leads to another. There is no alternative. Weak linkage, however, opens the door to contingency; developing individuals can choose the path forward. The basic freedom that Varela so passionately ascribed to organisms can be logically explained. For Gerhart and Kirschner, genes and cells are not in a relationship of cause and effect as a throttle might rev a motor. Cells interpret DNA in a kind of consensus procedure, Kirschner says, but they do not slavishly obey it. With this perspective Kirschner and Gerhart discover in the functioning of embryonic development exactly what Varela had predicted in his version of poetic ecology: cells and distinct tissues are autonomous to a certain degree. They act as a whole that interprets stimuli, making selective choices, and not as parts automatically obeying external laws. There are no such governing orders at the level of the body or cell. There are only signs with meaning. For Kirschner the information contained in DNA is not all-controlling instructions that the cell blindly follows; it is rather one stimulus among many. Other such stimuli include the composition of bordering cells, the temperature, the distribution of messenger substances and nutrients in the tissue, the degree of light or darkness in the surroundings, the overall health of the whole organism and its feeling of being alive.

Evidence to support this view is accumulating. The American biomathematician Gerard Odell and his coworkers, for instance, have made surprising discoveries that turn our understanding of the relation between DNA and metabolism upside down. They found that a certain set of genetic switches can be linked to one another only in one particular way to bring forth an organ. The process does not function through a genetic plan, however, but through a structural propensity. The organ proceeds to emerge even if the remaining genetic switches are randomly changed. When such a network is up and running, it is resilient against disturbances. It maintains its own balance. "The most simple functioning model was unexpectedly robust," Odell says. Today he is convinced: "Nature can design only this kind of networks" — ones whose structures are self-maintaining.[5] Organisms arise solely from organic networks that are animated from the bottom up through an orchestration of autonomous components, and not guided by a central control agency.

The absence of prefigured pathways leaves room for novelty to arise in embryonic development. The substitution of one single letter of the genetic code in a genetic switch can alter the critical mass of autocatalysis. This turns the dogma of mainstream biology on its head. Evolution in such a picture is basically possible because the cell does *not* obey the genes but interprets them freely, according to its particular circumstances. In this scenario, a change in code is not necessarily so influential that it can result in a lethal disturbance; in fact, the space for adaptation can lead to a different melody emerging from the genetic score. A meticulous blueprint of body parts and the sequence according to which they are constructed is not necessary because the tissues quite literally self-organize themselves. For a healthy body, development can reliably proceed if a few pioneering cells lead in the right direction. Complicated structures build themselves up of their own accord to form a coherent whole.

There are many examples of this basic paradigm of development. For instance, muscles automatically grow on bone structures, providing them with elastic connections no matter what shape the skeleton ultimately takes. Delicate veins emerge and extend themselves based solely on the amount of oxygen that developing tissues need. Growing flesh automatically receives sufficient access to all it requires within the organism to thrive and grow. Nerve cells form the right synapses through simple trial and error; they basically figure out the most useful ways to adapt to the situation they find themselves in. This profound flexibility in developmental processes means that what in one situation might result in an elongated finger among one of our remote ancestors might, in another circumstance, become the wing of a bat — independent of precise coordination and planning.

This astonishing internal coordination and harmony of growth processes stems from the developmental principle that every part of the body maintains active relations with all others. Organisms bring forth new ways of life without being programmed for it. They simply do it because each process possesses a critical degree of autonomy that at the same time always feeds back into the whole. If a change within one element of the overall framework of an organism takes place, not only is the particular module transformed, the whole organism as a totality changes. New forms that differ not just in a minor detail but in major body traits can arise with comparable ease. What has been a leg in an invertebrate can swiftly change into a part of the mouth in a new species which evolves from its ancestor, as indeed happened during the evolution of crustaceans. No tiny intermediate steps occur in this model, as the emerging whole self-constructs meaningfully around a renewed core. Embryonic development can be compared to musical improvisation. It needs only some inspirational input (genetic information) and then comes up with a meaningful performance. These findings also might help explain some of the historical "explosions" of biodiversity like that of the Cambrian period as found in the Canadian Burgess Shale, where within a short time span new unseen groups of animals emerged.

Effortlessness based on biological autonomy — many people find this idea just too difficult to accept. Those who do not want to give up the mechanical ideal of a divine watchmaker — especially believers in Intelligent Design — are particularly adamant in arguing that theories of evolution cannot explain the complexity of developmental processes. But the complexity of the eye or the brain, for example, may seem inexplicable only if you insist upon retaining the top-down, hierarchical perspective of central control. It seems impossible for highly complex developmental processes to have evolved on their own only if we deny the organism's autonomy, the subjectivity of every organic center that organizes the world according to its needs. This is a point that the champions of Intelligent Design simply refuse to accept. They cling to an ancient and rather Newtonian picture of creation that is even more wrong-headed than the genetic perspective already is. But classical genetic determinism partakes of the same traditionalist view that every organism must have a maker. Genetics simply shifts the source of this intelligence to genes.

But the deep biological reality, as many scientists are confirming today, reveals that the structures and functioning of organisms are not directed from anywhere outside. They come into being solely of their own accord as different units of organic matter mutually catalyze each other. We have forgotten for too long that the wisdom of the body is not something conferred from the outside, but something that manifests itself from within.

NERVES OF THE CELL, THOUGHTS OF NATURE

Some years ago, one winter I observed a group of squirrels through my large living room window. The animals quickly moved through the fir trees in front of the house. The ground was covered with a virgin layer of fresh white snow. The trees' needles were grayish green and the trunks dull red, far less shiny than the animals' fur. The squirrels skittered up and down the trees' limbs, then again clung tightly to the bark as if they were frozen, before continuing to chase one another in an undulating pace up and down the trees. There was one young squirrel, a delicate animal with nearly transparent red fur. The fierce chill (the thermometer had gone down to more than minus 20 degrees Celsius the preceding night) did not seem to matter. It raced after the others, leapt from one thin fir twig to the next higher one, oblivious to the world and at the same time anchored to the center of its universe. Its minuscule claws made faint scratching sounds on the bark that pierced the air like sounds that were themselves frozen and crystallized.

After a long while the squirrels finished their roller coaster play. They chewed off some pine cones attached to twigs, then sat on their back in a comfortable fork in the tree to nibble the seeds. They did not even bother to search for their hidden stores of seeds buried in the ground. The earth was too frozen to dig. Two years later, however, in spots where the small climbers had buried seeds, some slender seedlings sprouted. More recently, quite a number of young walnut trees had grown beneath the larger firs. The squirrels had slowly sown a forest. They were themselves a physical force!

The idea that subjectivity can affect physical reality is seen by most moderns as paranormal silliness. Feelings are one thing; hard physical matter is another. The subjectivity that inheres in the intricate assembly of matter we call an organism is usually regarded as separate and distinct from "nature." Orthodox science believes that only the most simple, inanimate material things can lie at the root of complex processes. We were all taught in school and college that every purpose, every individual aim, is the outcome of material chains of cause and effect — a belief that is still the guiding credo of modern biology.

But our connections to the living world as we experience them teach us the opposite. In the case of the playful squirrels we must concede that matter not only determines what life can achieve, but life itself is a powerful force shaping matter. They stand in a relationship of mutual dependency and mutual inspiration. They act on one another, as the great Canadian literary critic Northrop Frye liked to say, through "mutual interpenetration."[6] A living being, and hence a feeling subject, changes the configurations of matter in order to keep itself alive.

And these changes are not insignificant. The squirrels, for instance, plant trees and grow forests. Living beings in general are a geological force. The steady work of billions upon billions of beings transforms the surface of the Earth in the same way that body cells continually transform minerals and molecules into embodied selves. Think of the snails and crustaceans whose shells in the course of millions of years have sunk to the ocean floors and at some point in time have risen again to form hills and become the building material of Romanesque churches. What can we infer? The principle of causation, which works from the bottom up, is not the only one at play. There is also a principle of purpose acting from the top down. It acts by imagining a goal, by wishing to have a future. This purpose is not a matter of instructions and orders but of desires and interests. Many important things in this world are not explained by an analysis of how their constituent parts fit together to create a machine, but by understanding the origins of desires and how they express themselves and act upon the world. This dynamic is not found only in the world of humans. Purposes, the purposes of living beings to exist and to unfold, are a basic biological phenomenon. They are not sealed off from "nature," but on the contrary, purposes are an elemental physical power that is able to move and structure matter.

We find the ability to move autonomously and thus to arrange matter into desired configurations at all levels of biological existence. It is life's pervading principle. Organic subjectivity has countless faces. The proteins inside of cell plasma whose billion-fold connections crystallize into a vast interconnected meshwork that stabilizes itself, and the distributed computing network of genes, which with the aid of autonomous developmental modules brings forth the embryo, both provide scientific evidence for poetic ecology's idea of biological subjectivity. But it is not just the cell or embryonic tissues that are capable of such self-organizing, identity-forming behavior. Organisms consist of many layers of such networks, all of which show degrees of autonomous behavior. The nervous and immune systems are highly developed examples of meshworks generating a sense of self from the interplay of innumerable connections. The nervous system creates the body's perceived coherence — its sense of proprioception — and the immune system is permanently occupied with bringing forth a bodily identity and defending this identity against influences that cannot be incorporated into the organism.

All these networks are predominantly engaged with themselves and through this self-referential behavior generate subjectivity. Cognition researchers have observed that 98 percent of the nerve impulses generated within the brain and the spinal cord are not related to any processing of external stimuli. They are "talking among themselves." And like the biomolecules in a cell, nerve impulses are totally distributed and not separated into distinct cognitive containers. Neurobiologists can demonstrate that if a cat moves only one paw, its complete pattern of neural brain activity reorganizes and becomes totally different than the moment before. The bulk of neural activity reacts to the activity of other nerves, not to external sensual stimuli, whose salience nearly fades amidst the constant noise generated by the neural system's chatter. This continuous noise is nothing but the irrepressible self-expressiveness of life, the background murmur of subjectivity incessantly embracing itself.

A unitary living system reacts to any stimulus, and the whole living system changes to accommodate it, by creating a meaningful picture of itself within a dynamic world. Thus, the old image of the idea of the organism as a reflex machine, which has shaped our ideas of biology for a century, must be superseded by a new perspective. We must recognize the autonomy of living agents whose behavior does not blindly respond to a stimulus — the smallest discrete building block for a worldview of causality to which any complex system was to be reduced by a "serious" scientific approach. But the complexity that is constantly creating self-interested centers cannot be reduced. It is the truth established through an expansive, self-organizing creativity — the essence of what organisms do.

All the processes that lead to the self-construction of an individual within a larger environment are entangled in a most intricate manner. We therefore can see that the principles of an ecosystem are also at work in the body of a living being. As an organism is constituted by its cells, an ecosystem consists of relatively autonomous agents — its living beings — that are connected to each other in multilayered networks. An ecosystem can perform a kind of embodied cognition of its own. It, too, is a kind of organism.

It is surprising to see how the newly discovered role of the genes as switches and relay stations is similar to the workings of a nervous system. We could even say: In the same way as the genetic switches constitute the nervous system of a cell, living organisms are the nerve cells of an ecosystem. The oscillations of populations, the shifting frequencies of species and the changing beauty of a place can therefore be understood as a way of ecological cognition — the world seeing itself. In this sense, animals and plants, bacteria and fungi and all the rest of life are in a literal sense the thoughts of nature, as the Estonian embryologist Karl-Ernst von Baer called them. How a living dimension of nature fares — its worms and butterflies, birds and lizards, all of which are building up nature's quivering nerves and articulated sense organs — better expresses its health than conventional scientific measurements.

When we turn to a biology of subjects, we not only come to better understand many problematic areas in the life sciences. We also enter a bigger whole from which we have been estranged for a long time. We find that we ourselves and the inner laws of our biological functioning as well as of our experience are highly germane. We find them in the minute details as well as in the huge ocean of the bigger connections. The biosphere expresses something about us. In the next chapter I will explore why this is possible and what consequences this may engender.

When all the squirrels had fed, one after the other vanished. They climbed up into the tree crowns, running quickly along the outer branches and then jumping into neighboring trees. The scales of the pine cones made scattered dark patterns in the clear snow. Only the young squirrel lingered a little longer. It continued to chew peacefully. Then suddenly it started to run along an extended branch which pointed into the empty air. There was a willow tree maybe three yards away, but it did not seem very probable that the squirrel kid would be able to make such a gigantic leap. But it ran with full speed along the fir branch, slowed down near the tip, gazed down at the soft white snow — and then started over. It leapt and soared through the empty air in a long curve, hitting the snow with all four limbs stretched out. I remember the soft impression that its body and delicate legs left in the snow, even after it had dug itself out and vanished. It had obviously known that it would be a lot of fun.

Part Two: The Language of Feeling

In this section I want to shift our point of view from impression to expression. If we accept that to be alive is to be a self who cares for itself, and that this concern about oneself and one's effort to grow and prosper is first of all a bodily process, then it follows that the desire to unfold has many visible, tactile, audible, olfactory and other perceptual dimensions. Every organism's inner perspective as a self is simultaneously and necessarily an outward aesthetic reality. If a living being is not an insensate machine but rather is animated by values and meanings, then these qualities become observable. Meaning makes itself manifest in the body. The values that an organism follows are not abstract. They actually guide a body's development and coherence, whether the body is as complex as a human being or as small as a single cell. Feeling is never invisible; it takes shape and manifests as form everywhere in nature. Nature can, therefore, be viewed as feeling unfurled, a living reality in front of us and amidst us. This unity of experience and expression has become the focus of a new research field, affective neurosciences. It claims that embodied emotions are the prerequisite to understanding mind. The real enigma still lies in explaining how cellular world-making and existential experience are conjoined. But on this much, affective neurosciences agree: we can understand consciousness only if we understand the whole of life as emotional phenomenon.

CHAPTER 4:

World Inscape

You cannot watch the leaflets and flowerheads without knowing: You are related to them ... Springtime announces so vociferously that we too are a version of spring. For this is the reason for our delight in it.

— Lou Andreas-Salomé

A few years later I was in Estonia again. Kalevi had invited me to one of the legendary seminars he hosts for his students every year in summer. We met at the Baltic Sea at the tip of Puhtu peninsula. It was a tranquil, summery and serene moment out of time. There were no waves smashing the sea's glassy-smooth surface, which had the effect of expanding the sky, doubling its volume. The air was soft. Delicate milky and rose-colored streaks of cloud ambled slowly across a mild blue sky. The evenings remained full of light until very late, making it seem that the dusk would never end. Sea and sky dissolved into expanses of light. At the shore there was no maritime in-between zone of rocks sprayed with surf or salty, sodden flats. There were only wild rose bushes, still thick with flowers, which grew between massive granite boulders that aeons of water and ice had patiently crafted into crude spheres.

The chill and the blackness of winter had so thoroughly vanished that I could hardly imagine them ever returning. In summer, Estonia was a world of light. I could not imagine any other temperate region where wild herbs and spring and summer flowers erupted simultaneously, entangling each other in the ecstasy of a short summer season of dazzling brightness. The plants were showered with light for 20 hours a day by a sun that seemingly did not want to set. Summer in Estonia seemed to me like a paroxysm of passion that made everything possible.

The seminar took place in the former summer home of Baron Jakob von Uexküll, the iconoclastic biologist, who had constructed a modest residence at the tip of the peninsula. It was a spacious wooden house with a gable roof and a porch that opened onto a wide terrace. From there crumbling stairs led down into a neglected but romantic garden. Here we sat, in the shadow of rampant lilacs and old linden woods, discussing a new science of the living.

A couple of years before the Puhtu summer school, Jakob von Uexküll's son Thure had told me how he had spent the summers here in the house as a young child. I met Thure at his home in the southern German city of Freiburg, shortly before he died at a very old age. I remember the font size on his computer screen so much magnified that one sentence seemed to fill the whole monitor. We talked about a new vision of biology, but also about his father, Jakob, and Thure's early childhood at the Baltic shores. He had passed some of what he now remembered as endless summer holidays together with his father in the faded aristocratic setting on the Puhtu peninsula. Already then it must have felt like a bygone era that lingered like a mirage for a few years on the horizon before imploding during the destructive frenzy of World War I. A black and white photograph from 1915 shows a bright-looking blond boy by the hand of his father, who leans on a walking stick. By the time of Thure's boyhood holidays there, the weathered statue of Schiller — the very first in the world of many erected in the poet's honor — had sunk deeply into the ground in the Puhtu park. The beeches, the northernmost specimens of this European tree species, stood tall and commanding as if they had been there from the beginning of time. Putting aside his work on new scientific papers Jakob von Uexküll roamed the park and the seashore with his son, showing him the same realities of nature that his papers described in elaborate verbal accounts. Thure said that his father cherished one particular aim — to help young Thure learn to love plants and animals in their own right, as individual beings, and not as sentimental projections of his own emotions (the "sweet" wren, the "greedy" wasp).

Old Baron von Uexküll taught his son to relate the appearances of living beings to the factors that gave meanings to their behaviors and to their entire lives. Summer after summer, Thure underwent an intensive course in feelings, an intensive course for all his senses. There was the ceaseless sunlight, the light that still filtered through the blue linen curtains at 10:00 p.m. There were the grains of sand between his toes, the creaking floorboards in the house, the smell of wood fire and kelp, the cool breath of the trees, the joyous whimpering of the hounds, the melancholy tune of the bluethroat during the endless evenings, the warm milk before bedtime. The weeks in Estonia taught Thure how deeply entangled the inside and outside of living beings are. He learned how deeply any organism's feeling takes hold of its bodily appearance, whether it was self-conscious or oblivious of that fact. He learned to realize that very experience is saturated with joy or pain — and that any one of these feelings can be traced to a certain embodiment, a distinct bodily sensation and gesture.

ENACTING INWARDNESS

In the evening, after the first lectures, we set up a fire at the beach amidst the pebbles which were still warm from the sun. We sat on the boulders sipping cans of lukewarm Saku beer and inhaling the flavors of wood smoke and sweet, rotting seaweed. All over the sand flat, the rounded stones were strewn like the last outposts of reality before the lead-gray waterfront dissolved somewhere in the distance into the hazy atmosphere. In the bushes a great reed warbler repeated his signature song.

The light did not want to go away. It was late, ten or eleven, and still there was milky daylight. I got up, flipped open a new can of beer, and strolled down the beach. No air stirred in the breathless stillness. Could any place in southern Europe offer a more dramatic sunset? The light in these latitudes constantly promised something more to appear, an eternal becoming that assured us of some future fulfillment and yet eternally postponed it — a breathless tease. Russians had loved this sweetness when they still occupied the Baltic states. The former Estonian Soviet Republic offered the modest freedom of an unexpected resort holiday — the endless warm nights, loads of beer, maybe a summer love.

"Ah! When I hear your Baltic accent!" a Leningrad taxi driver once told an Estonian writer friend of mine without any bitterness at the course of history, "Then I must think back to the wonderful health cure I made on Saarema Island. The wide beaches, those sweet days." Estonia was the Soviet Adriatic, a summer playground full of fresh chances, hopes and beautiful melancholy.

I walked a long way that evening. Before turning back I wandered some distance into an old oak grove lining the coast. The trees' spreading branches had produced a spacious, airy canopy. The space beneath the trees was so vast that it felt like a lush pasture, a "wooded meadow," as the Estonians call it. This type of landscape used to be a commons grazing ground whose grass and acorns once fed the livestock. The coastal meadow was a searingly beautiful, magical scenery. It brought to mind the countless nature commons around the world, managed landscapes that balance the well-being of the ecosystem with the needs of its human partners.

Along the roadside, tall stands of Queen Anne's lace greeted me. By some strange accident its flowering umbel had been chipped off except for a tiny strip of tissue that held the blossom to the stalk. But the blossom, instead of falling over and withering, had managed to rally and stand upright with an elegant curve. I thought of Goethe, the poet and naturalist, whom von Uexküll often quoted enthusiastically. For Goethe, every being appeared in the form of a time-shape (*Zeitgestalt*).[1] Unlike the surface of a rock, the skin of an organism shows not only the passage of time and its physical effects, but also the plants reaction to the manifold changes it has endured. It displays how its life has been affected by various encounters and how it is predisposed to act in the future. The gestalt of an organism shows how it relates existentially to, say, the experience of a summer shower with its heavy drops of rain or to the sharp mandibles of a voracious beetle. The stalk of the Queen Anne's lace spiraled around its injury, searching for the sun. It was a gesture that reminded me of a fencer's movement, stopped in time by stroboscopic photography. The stalk showed the frozen sequence of motions which incorporated its past and its present, but which also displayed, in the upward direction of the plant's growth, its future. It seemed to me like a sculpture expressing the desire to be.

On that wooded meadow, I reflected that a plant is not only the result of all the past influences that have acted upon it. The evidence of these experiences is still there, still present. When a new stretch of bark grows around an abrasion caused by a carelessly installed piece of barbed-wire fence near a tree, we can witness the cause (the sharp tip), the effect (an injury to the plant), but also the *living experience* (the way the tree has coped with the wire by forming scar tissue around it, a protrusion of the bark). Organisms form scars over their injuries and in so doing bear witness to the original harm as well as to their self-healing. They embrace their past, making it the nucleus of their present. Their bodies *are* this past. But their visible, resolute growth also speaks to their future. Something immaterial seeks to express a necessary form. The achievement of this form, a gesture of the present, is a sign of aliveness.

I was thinking that if the physics of life can only be adequately expressed in terms of inwardness and feeling, this dimension necessarily must also be visible as a physical reality. If living beings realize themselves as selves by regulating the flow of matter through their identities, according to some deeply felt needs, then the matter that comprises an organism must express this subjectivity. It must display feeling. It must enact inwardness. It must *be* inwardness crystallized. We could even say that by embodying feeling, form allows us to feel the traces of lived subjectivity in other living beings. This subjectivity is different from our own in detail, but not in principle. If this is true, nature cannot be seen as the mute backdrop of an endless series of cause-and-effect reactions, but rather must be understood as a richly articulate, expressive medium. An organism's feeling, its inwardness, is accessible through its bodily presence.

That does not mean that other beings share and enunciate *our* way of feeling. This would be naïve. By talking about feeling, I do not intend to enlarge and extend the notion of a particularly human condition to all nature. Nor do I refer to the psychological unconscious. Feeling is much more basic and existential. It is an experience and a formative power that binds an organism together. It does not spring forth from the attractive and repulsive forces of atoms, but from a living unit's concern with its own persistence, growth and flourishing. Feeling means self-concern — and we do know that sensation. Feeling means inwardness, and it is this dimension that we share with other sentient beings, if only to a very small degree. Certainly the inwardness of many life forms is not very similar to human self-awareness and emotions, and to our sense of success and loss, grief and triumph. We cannot know to what degree we share these emotions with other beings. But we do know that we share the same trembling for our existence, its future, its unfolding, its flourishing. These are feelings that any autonomous living creature experiences. We are the same as all other creatures because we all have a vulnerable outside that mirrors our inward sense of vulnerability.

In the forest loneliness of the wooded meadow, the thought grew in me that the desire of the living I talked about in the previous chapters does not only manifest inside us. It also displays itself before us in the bodies of other beings. If we can best describe the physics of life as feeling, then this feeling nevertheless is always entangled with the matter the organism is made from. Feeling without matter is impossible. What we experience inwardly as emotion is something that happens outwardly to ourselves as bodies. And everything that happens to our bodies, and every way we react to that which happens, can be perceived externally. All life is a gift of matter. All living imagination stems from that unconditional surrender. It is matter that makes the potential of intimate, meaningful connections visible, which then makes our bodies more than "just" matter, but still, at the same time, nothing but matter.

For these reasons, poetic ecology, the new "subjective physics" of organisms, rejects the mainstream biology's obsession with explaining all qualities of organisms by atomic causal forces alone. In this respect, a poetic ecology is less materialistic than the mainstream biological thinking it wishes to correct. But at the same time it pays tribute to the body in its own right, in its role as the absolute ground zero for feeling, which is not possible outside the body. Hence, in a strange way, poetic ecology is more materialistic than many of the dogmas of contemporary biology, which echo the rationalist concepts of the past such as the idea that an abstract genetic code directs a body as it would a machine. The body, however, is not an abstract structure that works according to biological necessities alone; it is an irreducible entity of reality in its own right whose feelings of what is good or bad (with all the gradations in between) define its existence and activities. The physics of life identifies the unique individual body — and not the impersonal, generic anatomical machinery — as a crucial factor in understanding how living beings experience their worlds. The hidden feeling of aliveness is inscribed within matter — sensitive, delicate and highly destructible matter. The desire for life *relies* on matter. In this way it pervades and saturates matter with living significance and only thereby makes it beautiful. Life's subjectivity inescapably emerges into the visible, into the ecstasy of color and odor, of melody and touch. In this respect every living being is an open book, an individual looking glass through which the whole book of nature becomes readable. This transparency is not romantic projection. It is the result of our being inwardly the same as what the outward expressiveness of other beings shows. I will try to explain this connection in the next paragraphs.

Standard biology, still firmly anchored in the past, regards an organism as neutral machinery that cannot be understood from its outward appearance. Therefore, to be understood, it must be dissected. For the dismembering gaze of the classic scientific approach, corporeality as something meaningful in itself makes no sense at all. Here, the body can only be conceived of as a tool, not as a purpose in its own right, through which the ultimate, simple desire of be-coming real shines through. Poetic ecology, however, reconciles the science of life with the experience of being alive, and there-fore with what we see and what we love. From its viewpoint, as when gazing into the face of another human being, the whole of nature becomes a "thou."

It is startling to see the degree to which biology has abandoned interest in the expressive aspect of living beings. Aesthetics is not a part of its academic university curriculum even though all encoun-ters with other beings are inescapably aesthetic, mediated through the senses, possible only through sight and touch and smell and the beating of our hearts. Morphology, the venerable old science of living forms, leads a ghostly shadow life amidst dusty botanical collections and formaldehyde-scented repositories.

Yet many a biology student has been motivated by a deep fas-cination with living plants and animals. Nature first reveals itself to nearly all people as a source of liveliness and enthusiasm. Every child demonstrates this deep enthrallment when she plays on a meadow in spring. Immediately she starts to gather sticks and blos-soms, worms and beetles, and sinks deeply into the living world. Other peoples and epochs have taken this passion seriously. But not our modern, academic orthodoxy, which continues to regard any candid love for nature as a throwback to romanticism and its questionable cultural heritage. It is worth recalling that romanti-cism tried to restore our original relationship to living things, a task that it never wholly accomplished.

Tribal peoples nearly everywhere live in a relationship of deep reverence and awe in respect to animals, plants and even rocks and lakes. The Australian Aborigines narrate the dramas of their my-thology based on natural beings and places. The force of creation speaks to them with and through the voices of other creatures, as well as those of their own bodies. The senses provide access to the deep knowledge inherent in the ecosystem of which humans are a part. An Aborigine elder explained that he experiences the power of sacred paintings like the "fresh sparkling of water." Their power exerts a force on the observer; it immerses us corporeally and psychically, like a bath in a cool eddy of a wild brook.

The magical force of beginnings can be sensed everywhere in the ordinary forces of nature. For this reason the time of creation, the Aborigine's "Dreamtime," can be expressed only through real and ordinary beings and landscapes. In their ritual art, the Aborigines, therefore, use natural materials and refer to the sign language of biological reality to partake in the original forces of creation. We do not have to travel to Australia to witness similar worldviews that could inspire a correction of our own. In Europe, in the Middle Ages, when earthly realities generally meant less and heavenly promises everything, nature was still experienced as revelation. It was widely regarded as the "second book" of God, which the creator had written in living letters so that everyone who had a body could read its meanings.

Mainstream science, on the other hand, is so obsessed by pure facts that its image of life sometimes seems as if an art critic would attempt to describe a painting only by listing its color pigments and their wavelengths. Science's hyper-literalism misses much of what is really going on. Not so long ago, even philosophers took their lived experiences more seriously than they do today. Immanuel Kant considered the manifoldness of life forms and the question, "What is beauty?" as inseparable. He regarded them as different ways of looking at the same issue.

On my long evening stroll along Estonia's flowering coast, modern science's success story of building on a renunciation of lived experience, of the world as it is, seemed to me like a lengthy and painful detour from what really matters. I realized that with our craving to build a new and better world we have thoughtlessly given up that one crucial sphere to which we are linked by the umbilical cord of life. We have attempted to sneak away from our "Siamese connection with all other beings," as the novelist Herman Melville beautifully described our situation. We have tried to escape from ourselves.

Poetic ecology reverses the current viewpoint. We do not experience ourselves as subjects solely because we are endowed with a mind. Rather we have a mind because all life is subjectivity. Our subjectivity stems from a desire without which any physics of life will remain incomplete.

THE REBIRTH OF THE PLANT'S SOUL

As we realize that subjectivity is all-pervasive, we can serenely extend the idea of universal feeling from animals, seemingly closer to us, to plants. Even if plants, compared to animals, respond to stimuli slowly and lethargically (they do not have nerves, as animals do), their lives are governed by the same striving for plenitude and wholeness as ours. And biologists are increasingly discovering how energetically green organisms show their zest for life. The signs are everywhere. We just haven't bothered to pay attention.

Instead of electrical impulses, many plant cells communicate through messenger molecules. These are particles that float through the tiny vertical tubes in stalks and stems where plants' ubiquitous body liquids circulate. These particles can also exit the leaves as gases. In this way, vegetation can even issue cries for help. A few seconds after a plant's stalk is injured, the root tip deep in the soil is flooded by a cascade of alarm molecules. Plants are capable of displaying all the qualities that have so far been reserved for animals. The chief difference is that vegetation is slower, more sedate and silent. It is as if plants lead lives in another dimension that is barely accessible to our senses. But they are subject to the same existential challenges we have to face.

Plants even sense touch. A gentle stroke suffices to make them shift their direction of growth. And plants can see — not with animal eyes, but with their whole bodies. Their green surface is one giant organ of vision. Because their cells are able to discern the proportions of different wavelengths in the sunlight, they can change growth in the direction of best energetic yield. Other senses also allow vegetation to enhance its self-flourishing. Seedlings taste nutrients with their root tips. Spruce sprouts lodged in a steep rock crevice can feel the pull of gravity through minuscule crystals in special cells that act like a gyrosensor in a handheld device. Little corn sprouts use their body surfaces to smell alarm substances emitted by neighboring plants that have been invaded by parasites.

All these discoveries made by curious botanists (which often have been rediscoveries of knowledge from a time when we had more respect for the capabilities of our green fellow creatures) are a source of constant suspicion for some scientists and a spring of delight for others. What seems to be the astonishing skills of plants is indeed not much more than the basic autonomy found in every living being, and therefore, something that every organism masters without too much effort. The intelligence of plants reveals life's ubiquitous intelligence. What we observe in humble vegetables is life itself, which longs for continuity and, therefore, consolidates itself around this subjective concern. The plant that directs its growth tendency to the light (a tropism) does not understand the arithmetic of wavelengths; it simply perceives light as good in the form of a positive affection. This is not so different from those of us who prefer to sit neither in the direct sun nor in total shade, and therefore, without thinking move to the most pleasing location.

Today's botanists have used ingenious experiments to confirm the subjectivity of plants. The plant physiologist Anthony Trewawas of Edinburgh University has observed that identical plant clones — multiple vegetative twins whose DNA sequences are identical to the letter — behave differently, even though room temperature and substrate moisture are the same. They are clones, but their bodies unfold into individual shapes. "They individually choose between different options," Trewawas says.[2] Every sprout has its own preferences. Each is an individual, committed to its ongoing existence in its particular body and not simply an automaton carrying out a genetic blueprint. For Trewawas it is just this stubbornness that must be viewed as intelligence. Intelligence, according to the meaning of the Latin verb *intelligere*, means to be in between, to be able to choose. It signifies the ability to make a decision, and hence the judgment of a distinct self for whom a choice means something — survival, growth, flourishing. In this sense intelligence and life are one and the same thing.

METAMORPHOSES

In plants' bodies, whose presence I sensed so acutely on that long Estonian evening, I can feel powers at work that also are active in myself. These are the powers of life. We do not only subjectively feel them; we objectively see them in other bodies as well, although this objectivity is not absolute but poetic. Conversely, subjectivity, as it arises within our bodies in space, is entirely empirical. It exists. It can be observed. Empirical subjectivity is experience, which leads to poetic objectivity in its expression through the body itself. This is a wholly materialist approach to biological phenomena but one that makes freedom not only possible but necessary. Its empirical basis lets us glimpse a world beyond our familiar mental template of cause and effect. In all our clumsy tenderness, we living beings are instruments of an inner dimension that, when seen by other beings, can be understood.

In the blossom of the flower, the bliss of beginning is real — and at the same time becomes a metaphor for my individual life. The blossom is a primordial metaphor I generate with my body and which I cannot do without. It enables me to understand myself — to see my own beginning, my own hoping as a facet of the general principles of the living world. I partake in a universal condition which pervades all living forms and which makes life possible. We could call it the *conditio vitae* in contrast to the *conditio humana* that existential philosophers like Karl Jaspers coined to describe the profound inner abyss of a human life. Seen from the perspective of the *conditio humana* we are a suffering mankind. We are alone and thrown into the freedom to choose. Observed from the angle of the *conditio vitae*, however, we are not isolated but deeply intertwined with the biosphere, not only by our bodily metabolisms but also through the existential meanings and needs that all sentient beings have. We are brothers and sisters of everything alive and feeling. We are embraced by that which again and again comes to life, and through this we can transcend the inevitable, constant suffering of life.

If we look carefully, we can observe that the embodied experience of nature has shaped the form of our language and our verbal expressions throughout the history of our species. We still articulate our emotions in elemental metaphors: love is hot, anger steams, wrath rages like thunder, the soul thirsts. And vice-versa: our experience of the elements is also a passionate act. There is the tender air of springtime, the furious thunderstorm, the sweetly fragrant, yet prickly rose. The whole biosphere lends itself to emotional resonance and becomes a symbolically extended self. This is not the case simply because cultural habits form our language, as is generally assumed today. We create linguistic expressions on the basis of embodied experiences, and we can find the reference point for these experiences in other beings that embody the same existential feelings.

Symbols thus become the language of the psyche. In this sense, the vegetative world in front of our eyes is nothing other than our own inwardness. Nature shows this inwardness as a silent mirror, expressing before us what is unnameable because it is within us. In contemplating the life of a tree, a human can comprehend similar forces that he knows exists within himself without having to analyze them, because, again, they are part of the self with which he organizes experience. The knotty oak and the sea of grass caressed by the wind are living reservoirs of understanding, as when I sense my own powers of endurance or joy or melancholy. Because the plant is always growing, it illustrates not only the principle of self-conservation in life but also the intensification of self. The plant is aliveness made physical, embodied in lush leaves and buds impatient to bloom.

The day before the summer school began, I stopped at the side of a small country road on the way to Puhtu. I stood before an imperial linden tree, which my map identified as a natural monument. Eight branches emerged from a single massive trunk like fingers splayed in a wild gesture. I guessed they must have been rising upwards in this posture for centuries. The branches tossed their leaves into the air like the copious jets of a fountain, spewing out green cascades speckled with tiny blooming stars aglow with fragrance. The tree was not only a resplendent spectacle. Its fissured branches were covered with lichen and moss and, to my surprise, decorated with flickering colorful ribbons and a new silk scarf that someone had obviously tied around a branch just a few days earlier. I suddenly realized: this is a holy tree! My musings were interrupted from time to time by a rickety old Pobeda or Moskvitch automobile passing by, kicking up a cloud of street dust.

There is no culture in the world in which trees are not worshipped as symbols of renewal — trees like the Persian Haoma, whose sap bestows eternal life; the Chinese tree of life that reaches 100 thousand yards high; the Buddhist tree of wisdom, from whose four main branches spring the great rivers of life; Yggdrasil, the holy ash tree of the Norsemen, which holds the earth in a balance between the underworld and the higher regions by means of its trunk and its roots; Apollo's laurel; Aphrodite's myrtle. It is no wonder that trees were chosen to signify that which endures. Sacred plants in archaic cultures exhibit in themselves the cosmic progression of youth and propagation, aging and rebirth. Plants have become symbols that express the feelings of everything that is alive, creative and fertile and which must die. They are the naked avatars of carnal existence.

Unlike we self-conscious creatures of the animal world, plants do not know the shame of physiological intimacy. Their bodies have no functional inside. They are nothing but surface. They have no inner organs, only exterior ones. Flowers are naked in a much more profound and literally comprehensive sense than an animal could ever be. In contrast to all animals, which carry the bulk of their crucial organs enfolded within their bodies, plants have unfurled their functions to the outside, so that everything becomes visible — their respiration in the transparent green veils of the leaves, the history of their growth in the rising cascades of branches and twigs and, most unashamedly, their unbridled sexuality in the overflowing ocean of blossoms, seeds and fruits. From an unemotional standpoint a plant is a single oversized sexual organ. Apart from the narrow tubes that transport water and nutrients, they conceal hardly anything. Plants are inwardness enthusiastically offered up to the world. The feelings they present to other creatures might not consist of an inside as intensely differentiated as it is in animals, but on the other hand, the plant subject makes itself totally visible, palpable — and even readily edible. If the diversity of impressions that a flower conveys might be limited, the luxuriance of its expression is all the more intense.

THE PSYCHE IS A PLANT

Because plants are so outspoken in enacting the drama of their existence, they exhibit all the aspects of being alive which our own bodies obstruct from our view. We basically know our bodies merely from the outside, as physical forms — from our gaze that wanders from our hands to our toes, from a quick look in the mirror, from touching our skin. Even the mouth, ears, anus and vagina are but entrances into an outside that just happens to be folded into the body. Our inner organs and how they work are more or less imperceptible. The normal tensions of our body related to being alive, the tension of our muscles, the tautness of our flesh, are only noticed when we strain them. Normally, an individual's essential life functions remain subliminal or invisible. The liver's action, the transforming labor of this most important of all metabolic organs, is entirely hidden from consciousness and sight. We can notice what the liver does, however, in the metamorphotic character of a plant that is continuously converting matter and light into a substantial body. That is why a plant can be regarded as our ladder down into the realm of the unconscious beneath all explicit cognition. The sight of botanical life processes, which present themselves with such open abandon, informs our own experiences as subjects. We can look to plants when we talk about our souls.

One could be tempted to say that a plant's realm is the exact opposite of our inner lives. Unlike us, the vegetative world consists of pure physical bodies expressing the desire of living beings. But it would not be accurate to draw a strict line of demarcation here: the vegetative dimension includes us. Brain researchers estimate that unconscious cognitive acts account for 98 percent of human neural activity. For the psychologist C. G. Jung, the unconscious is not some virtual space inside or a form of ungraspable consciousness leading a purely mental existence separate and apart from the body and its swelling tides, as some people still tend to believe today. For Jung the unconscious is the body. As seen through the lens of a poetic ecology, this idea sounds uncannily contemporary. What thinkers like Jung interpreted as the unconscious, the new biology regards as the intrinsic subjectivity of the body that permeates the simplest cells. The body unconsciously knows through acting what is good for its continued existence and what could be devastating. The body is able to show the suffering of the self, as becomes most acute in severe psychosomatic diseases like anorexia nervosa, or autoallergic sufferings like asthma or multiple sclerosis. But this expressiveness is a dimension that we can never grasp directly; we can only access it as it is mirrored by our moods and symbolized by our felt desires and needs. We can only see it in the same way we encounter the life of vegetation.

On this silvery evening all these ideas seemed self-evident to me. It was so obvious that life yearns to satiate every fissure and cranny of the cosmos with meaning, to insinuate all its creative passions into the world as a jeweller might fill an elaborate mold with gold. The insight hit me like a blow. If an organism's feeling is revealed by its outside, then the whole of nature must be understood as one huge interior — a space of meaning, a topography of inwardness, which is experienced as an outside. The mystery lay before me, and it was unfathomable — and yet at the same time as accessible as bright sunshine.

The external world, thus experienced in a mode of necessary imaginative connection, does not remain something which is only extended and material. It becomes a topography of meaning. To describe this interior dimension of lived reality, poets, not biologists, have invented the most apt terms. The poet Rainer Maria Rilke proposed the German designation *Weltinnenraum*, which could be translated as "interior space of the world," or "world inscape." *Inscape* is a concept another poet, Gerald Manley Hopkins, used in the second half of the 19th century. For Hopkins, a deeply sensitive Jesuit priest, inscape was that which had form and meaningful expression at the same time. The *world inscape* is that dimension of the world which is not only spatial, but also filled with felt meaning. This space is the realm of poetry — but also of poiesis, of creation. It is the realm of organisms. They are all connected by the *conditio vitae*, which is shared by everything alive. What unifies all experiences that can be expressed in this deep dimension is their felt value for the organism that wants to sustain and unfold itself. These values are not material, nor are they a set of optimal system parameters for life. They are the degree zero of all meaningful behavior, the desire to have a future. This desire is something symbolic, something that occupies no space but which is nonetheless able to move the heaviest weights with ease. The value of any encounter is its meaning for intensifying and prolonging the coherence of the self.

Over the centuries, humans have again and again made the mistake of seeing only one of these two perspectives, the inside or outside, at the expense of the other. For today's science there is only outside. Science pursues an ideal of mapping the outside onto everything inside, reducing all things to purely exterior physical relationships, including what we subjectively experience as inside. In the academic tradition of the humanities, by contrast, only the interior counts. In its view, even matter, or at least how we perceive it, is brought forth by interior mental acts, by social customs or by our language. This bias in favor of the mental and abstract sphere has a long tradition. Even the big monotheistic world religions take a one-sided stance when they claim that in the beginning and in the end there is only logos, the word. They firmly embrace a dualistic picture and banish embodied experiences even though mystics at all times have tried to make space for body, feeling and the experience of living unity — a quest for which not a few of them had to pay with their lives.

How different religion would look if it did not accept this separation of inside from outside and did not assume that God created the world as an artisan creates an object. How much better to see the divine as transforming its essence into an unfolding, living cosmos that is the incarnation of subjectivity-becoming-matter — a subjectivity that needs the physical body and the joy and the suffering of innumerable beings in order to experience itself.

In the contemporary worldview, we rarely try to integrate or reconcile these two extreme standpoints. We basically accept the deadlock of mind against matter. We consent to combine them in a technically and economically useful way, but also in an emotionally deadening manner. Inside is "us"; outside is the whole messy remainder of all that is; and each is separate. This worldview only became possible because we agreed to split the world into two opposing dominions that never overlap and whose delicate equilibrium is ignored. By compartmentalizing the world in this way, scientists are granted a license to do as they please without needing to account for lived experience as both body-mind and feeling-matter. They do not have to acknowledge empirical anomalies and inconsistencies because the stipulated ontological split of body-mind preemptively renders such complications moot. When the focus is purely on the relationships between material particles, external relationships are all that is seen; the interior, the psyche, is invisible and inconsequential. Thanks to an a priori agreement about ontological terms, separating body and mind and matter and subjectivity, manipulation and value-free research and experimentation can roam free. Correspondingly, the specialists of the human interior, from social scientists to artists to poets, have forfeited any authority to judge what goes on in the realm of "pure matter." And everyone is happy.

The non-scientists and humanists seem to regard this amicable division of the world as the best that can be done. A great many of them have accepted the scientific worldview that life is bereft of meaning, values and goals. They have convinced themselves that, in any case, we are trapped in the mirages of our language and cannot make relevant statements about reality. The rejection of the material world which produces meaning on its own accord reaches such a radical extreme that many linguists today believe that we can experience a feeling only if we first know a term for it. We can feel only what the language game allows us to express. Unfortunately this is never able, many think, to reveal anything about reality.

All this leaves out existential experience, which can only happen within living, sensitive flesh. Such experience is not arbitrary or conjectural; moreover, it can be shared with any other living being. How strange that this realm of empirical phenomena — the delicate breathing web of nature — is an orphan subject that neither of the two scientific cultures of our world, the hard sciences and the humanities, can comprehend with their brands of rationality. Still, the truth of life is there. It is not difficult to find because we carry it within ourselves, in an inescapable middle ground — inescapable, because matter is not able to be real without feeling.

The German biologist Gustav Theodor Fechner, who held an influential position at Dresden University in the 19th century, had a beautiful way of describing the dialectics of inside and outside. For Fechner the world has both a dayside and a nightside. Surprisingly, according to Fechner, the dayside is not the reality of our physical bodies moving about in the world as physics describes it but our own inner experiences and perspectives. Fechner considered the scrim of abstract science as the nightside — the dark exterior which obstructs our perception of reality. For him, only our inner experience of our material bodies and our feeling of self are real, and they are the only way to truly understand the character of the world. The logic is quite simple. Organisms are inextricably a part of this integrated reality, which itself is experienced from the inside and outside at once. Only in us living organisms is the full character of reality completed. Only here, in the expressive meanings of a material body can that which exists experience itself as such.

I had left the forest and returned back into the open, walking in a long arc. I had to pause in awe for a few moments. On the pastures, innumerable white anemones, their hairy stalks receding into the distance, were looking across the sea toward the horizon where the sun had set. The light seemed like it did not want to go. Everything shimmered softly, a last fluorescence that stubbornly hovered in the air. Cove after cove was immersed in the white foam of the anemones. At the furthermost tip of land before it sank into the sea, the Earth had cloaked itself with beauty beyond measure.

The next day I woke up early, well before six, and the landscape was already inundated by full daylight, pouring through the blue curtains. The warm room smelled of old wood and linoleum flooring. I got up quickly and walked on the narrow path to the beach. The water lay as still as if it was a solid surface. When I waded into it, the chill at first took my breath away, but it became more tolerable when I had done some strokes. My hands dipping into the water sent out a small wave, splitting the liquid mirror and giving it a structure and momentum. The process reminded me of the physics theory that claims a first, small asymmetry in reality triggered the Big Bang and gave shape to the cosmos. The sky's brilliant orange and blue flared between my fingers. I turned back my head. The boulders on the shore reclined across the landscape like an endless ellipsis.

CHAPTER 5:

Affective Neuroscience: Mind as Symbol of the Body

We can say that a mental state is conscious if it has a qualitative feel — an associated quality of experience. These qualitative feels are also known as phenomenal qualities, or qualia for short. The problem of explaining these phenomenal qualities is just the problem of explaining consciousness. This is the really hard part of the mind-body problem.

— David Chalmers

A few weeks after the beginning of my biology education I participated in an unforgettable experiment. It happened in an introductory course in which we studied the different major groups of organisms for several hours every day. Bacteria and single-celled algae lay behind us. That afternoon a behemoth of the unicellular, protozoan domain was on the schedule — paramecium, also called "slipper animalcule," for the elongated oval shape of its huge single cell.

The problem with observing the tiny slippers was the fact that they swim way too fast. I had labored a while to fix the cover slip on the glass slide with four blobs of Plasticine on the corners, and underneath I had injected a bead of the liquid brimming with protist-rich life. The murky broth filled the microscope's view. I focused and saw ... spinning shadows. In preparation for the experiment that was the lesson for the day, the teacher had put two tiny brown glass bottles on the laboratory table, each with a syrupy liquid sticking to the inside. When he saw my searching gaze, he handed me one of them. The label read "Proto-Slow."

And behold — immediately after I slipped a drop of it into my tiny aquarium the flying protozoans slowed down in a dramatic way. It was like a *Star Trek* episode where a starship suddenly goes out of warp speed down to zero and for a moment every movement is in slow motion. It was as if the slipper animals had taken a psychedelic drug. Their tiny body hairs flickered more and more slowly. As the elegance of their motion had coagulated into a laborious crawl, I could now study the details of their bodies. I could also draw them now, as every student was supposed to do.

After 15 minutes of awkward gazing through the ocular, I had to move to the next stage of the experiment. To prepare for this, the lecturer passed around the second vial. It contained an acid that was meant to freeze the trance-like dancing under the cover slip. The aqueous solution was a deadly toxin. The dying animalcules would shed the delicate hairs and lashes that so lavishly covered their bodies and show us students the complex filigree of their construction. They were built from only one plasma chamber, instead of trillions like us multicellular beings, but in some respects they seemed more complex than other microscopic life forms with more cells. After all, they were moving and beautiful and huge. Undaunted, I took the pipette and dribbled a double dose of the acid beneath the cover slip.

What I saw next took my breath away. The flock of slipper animalcules were besieged by tiny waves of the approaching acid. Like a desperate army trying to flee an invisible enemy, they wanted to escape but could not. The slipper animals were slow, so slow. It was a doomed escape. One after the other was engulfed by the toxic substance. And when they were hit, the tiny beings writhed and doubled up as if their bodies had been scalded by boiling water. Involuntarily I hunched my shoulders up in alarm, writhing over the microscope until the struggle was over. And indeed, the paramecia had shed all their body hairs and remained nude and motionless as stranded wreckage. As promised, I could perfectly observe the minute structures of their carcasses. They did not move a micrometer. They had become sculptures of death. Even more so — they *were* death that had taken a particular shape.

WE ARE OF THE SAME STUFF

In poetic ecology we find the dimensions of inwardness and of feeling everywhere. Every event has a meaning for the living coherence of the body. Each contact with the world is an emotive act. Feeling begins with the expression of a body that appears as a mirror of what has been inflicted upon it. The feeling materializes regardless of whether the sentient being "knows" what is happening to it or not.

In these reflections lies the key to how our own subjectivity is related to the world. This raises questions that continue to haunt contemporary brain researchers, questions such as the one American neuroscientist David Chalmers expresses in the epigraph for this chapter. What connects living feeling with consciousness? How is consciousness built up from feeling? Who has it? Only humans? Higher animals as well? Or all beings? If feeling is the most important force holding a body together, then the track that leads to consciousness lies here and nowhere else.

Poetic ecology for the time being does not raise the question of consciousness. This seems meaningless as long as most researchers ignore the fact that organisms must be considered as subjects with sensations and selves. Our view, which is fixated on the human mind, misses exactly this character of biological individuality. Hence the unsolvable riddle: "How does the mind get into the machine?" For a biology that so firmly believes that the mechanics of life is the only reality, there is indeed something spooky about the plenitude of experience, inwardness and consciousness, which the genus *Homo* has so abruptly introduced into natural history.

But a new world of insight comes into view if we consider that the "mute body" embodies and expresses meaning. Then, suddenly, exterior forms become as important as consciousness. Then the body already expresses the standpoint of a self. The body chooses and interprets and, therefore, enjoys that fundamental freedom, which for a long time we only accorded to self-conscious beings, namely, humans. Because of all this, there is no reason to search for two different reasons for a behavior, one physical and one psychological. Behavior is one unified biological process. Consciousness is one form of experience expressing life's desire for continuity and unfolding. It is this experience as manifested from the inside. Consciousness, however, is not necessarily superior to the silent struggles to flourish that are also visible in other beings; it is another aspect of the same phenomena.

Some years ago, when I was writing my doctoral thesis in Paris, I made a moving observation while riding an 89 line bus through the fifth arrondissement. At the Panthéon a young mother with a sleeping child in her arm boarded and sat down on the bench in front of me. The baby lay snuggled against her neck, arms dangling down his mother's back. He moved his fingers in his sleep, straightened them and then closed his fists with a slight quiver. The infant slept. He was not conscious. And yet in this fleeting moment as I watched him on the bus, the infant was totally present in his feeling, as present to himself as any being I had ever seen.

I suddenly thought, as the bus worked its way down the Rue de Vaugirard, that consciousness is not the most interesting point about life. It is only one dimension among many others in which organisms experience feeling. The reactions that are visible to the outside world might even be more intense when the central nervous system plays a subordinate role. Perhaps the aesthetic plenitude of unconscious gestures and developmental patterns that shape the body may be even more eloquent and expressive than any conscious account. In poetic ecology, what matters most for all organisms is not conscious behavior, but acting in a way that makes sense. Any metabolic activity that keeps an organism alive makes sense. What makes sense can be totally unconscious — and, as we have seen, most of a brain's activity is hidden from consciousness. For this reason, the question, "What can he feel?" when looking at the face of animals and even unconscious humans is misplaced. We cannot know.

But we can feel. So we should ask instead, "What does this situation *mean* for that subject? What existential value does it convey? What is the feeling?" We can infer answers to these questions from a host of embodied signs. But to understand them, I must draw upon my own knowledge as a sentient being, recognizing other life forms as made from the same stuff as me. The stroke patient who cannot speak anymore, the slipper animalcule under the cover slip that doubles up and dies when hit by waves of picric acid, the bald hens in their stiflingly hot hangars, the unwatered plant withering away in silent despair — they all unconsciously show what they feel is happening to them.

The feeling of these beings is so unmistakably revealed in their gestures that we cannot help but show empathetic reflexes and compassion for the familiar, common experience of suffering. Indeed, we are totally unable to steel ourselves against such feelings because we know the same gestures from within ourselves. We know what we sense when our body bends in pain. We can summon up every speechless expression through our own feelings even if we have never ourselves had the experience we witness. We are able to empathize with other beings only because we are all alive. The principles of our body, which we share with all other organisms, form a universal translation device for the stirrings of life, for its aching traumas, for its happiness. Anyone who pretends not to be emotionally sucked in when another being shows its feeling, is dissociated from himself. Anyone who ignores the suffering of other beings who may not have the language, intelligence or personal consciousness that we do, not only despises those others but betrays himself. To deny the reality of suffering, he deceives his ego with the illusion that we humans are different.

Life itself is feeling. Consciousness is only intensified feeling, reflexivity becoming reflexive to itself, always mirroring the basic feeling of life. There is no subjectivity decoupled from the body. This is no longer an esoteric stance in biology; it has become the only view from which certain questions can be approached. For instance, only in accepting that the body is a feeling self can a famous quandary in brain research be resolved, the so-called Libet experiment. Neural researcher Benjamin Libet observed that if a subject raised his arm, the muscles started contracting some fractions of a second before the decision was "made" in the conscious part of the brain. This seems to prove that consciousness as a controlling force is bogus; it is simply a pleasant veneer on a basically deterministic machine. The Libet experiment has sparked a fierce debate that is still going on today. But the determinist interpretation of the experiment is based on the wrong assumption — namely, that the body is a machine. But if the body is recognized as the autonomous subject, then any kind of consciousness is the symbolic expression of that fact, an aspect and manifestation of the body. No wonder the conscious experience of this comes later. The basic autonomy of embodied selfhood has already exerted its freedom of choice.

BEYOND THE MIND-BODY PROBLEM

Biology need take only a small step to incorporate the reality of feeling into its official doctrines. In brain research this process has already begun. Neurobiologists think that the foundation for human consciousness lies in our subjective emotions. In particular, Antonio Damasio, a brain researcher at the University of California at Los Angeles, has shown through a long series of experiments how bodily sensations influence the mind. The interaction is so strong that in most cases when the body's natural tendencies are prevented from developing, mental development is endangered as well. For Damasio, emotions are mediators between body and consciousness. They represent how well the whole embodied individual is faring.

From this perspective Damasio and his French collaborator David Rudrauf created the self-representational theory of consciousness. They believe mind to be a kind of virtual enactment of the whole of the body. If feeling can be regarded as the existential meaning of what happens to an embodied being, then consciousness is also such a symbolic mirror. It is a second-order mirror — clearer and more precise — but at the same time far more remote from what it signifies. It follows from such a view that mind has little to do with logical symbol processing, contrary to what researchers have long believed. The roots of mental phenomena rest in the body's subjectivity. "The attempt to explain consciousness without taking into account subjectivity and emotion is like trying to understand the blood circulation without considering the heartbeat," Rudrauf and Damasio maintain.[1] Admittedly, the two scholars restrict their conclusions to the human neocortex. They still suppose that subjectivity is solely *our* business, a feature of human beings alone. Anything more would still be regarded by the neuroscientific establishment as esoteric speculation.

AFFECTIVE NEUROSCIENCE: LAUGHING RATS

Another researcher goes a decisive step further. Jaak Panksepp has been researching the brain for decades. Although he works in the US, Panksepp originally is Estonian. (Who should wonder?) Panksepp is developing what he calls an affective neuroscience, which is meant to encompass all biological species, not only humans. Panksepp uses laboratory neuroscience to track the emotional capabilities and habits of animals. "If we really want to comprehend the motives governing the behavior of animals and man we must understand their emotions," Panksepp says. He aspires to achieve one of the "most fascinating breakthroughs in brain research of the new century." It might be *the* breakthrough.[2]

Panksepp has long been convinced that other animals feel the fundamental value of existence as we do and that this subjectivity is not merely reflexive in the manner of hunger, thirst or fatigue. Many animals that we never would have imagined experience emotions. Panksepp arrives at this conclusion because he is one of the few neurobiologists who designs experiments that actually seek to elicit laughter from animals. In one setting he playfully tickled the bellies of young rats while tracking their brains' electrical activity and using ultrasonic microphones to record their voices, which are mostly inaudible to human ears. Panksepp discovered that we've never heard rats laugh because they giggle in frequencies that are too high for our ears. Rodents do not behave differently from a toddler who is seriously tickled and squeals with pleasure. And the rats' enthusiastic chirp is related to the same neural vibration that signals pleasure in human brains.

In another experiment, Panksepp turned rats into morphine addicts. He observed that rats suffering from arthritic pain readily resort to opiates when they are offered to them with food. Opiates' soothing effects work in those regions of the brain that, in earlier stages of evolution, had already developed in the central nervous system of archaic reptiles. The processes going on in the brains of crocodiles and rats are the same changes that happen in ours, causing emotional experiences. Panksepp supports this idea with further arguments based on further experiments. When researchers stimulate the areas of human brains that correspond to the same brain areas in other vertebrates, both show exactly the same responses. Why shouldn't the emotions that mediate this behavior be the same in both instances? If a certain region of the brain is aroused, human test persons break loose with laughter — and young rats start their cheerful chirping if the corresponding zone in their brain is stimulated. In all vertebrate animals, the deeper and evolutionarily older regions of the brain use the same regulatory circuits and messenger substances. In terms of developmental genetics, researchers have found that the same genetic switches that guide the formation of our brain are active in flies and crabs. Panksepp concludes, "The common belief that only the neocortex of our big brains is able to produce core affects is as naïve as the geocentric world view."[3]

As we have seen in the last chapter, even a cell experiences value in a way that guides its sense of self-preservation. The value of sustaining one's own embodied existence is, after all, the pacemaker of life. Feeling, which is the experience of this value, is the most fundamental level of reality for an organism. Feelings show how close any action comes to fulfilling that core value of all life, the value of continued existence. Feelings are a subject's eyes. This subjective authority, this striving selfhood, produces feelings that direct the entire being; Panksepp calls it the core self. The status of the body is mirrored in the core self. The core self is where the meaning of bodily processes and their interaction with the environment are collected and interpreted.

The core self, nonetheless, is not a control center; it is a subjective viewpoint. The assessment of bodily states is not something that can be located at a particular place in the material architecture of brain anatomy and chemistry. It has no spatial dimension. It does not consist of matter. The feeling of one's own embodied persistence and drive is achieved in a formal, not a material, way. It is achieved through the translation of meaning into meaningful gestures, through a "telling form" that resembles a work of art. For an organism to make its own situation transparent, there are neurosymbolic processes at work. In Panksepp's eyes, the core self values its experiences and inner states by making them appear as emotion. To put it the other way around, feeling is what bodily processes *mean* for a subject. Through feeling, the unconscious signification of everything that happens in metabolism translates itself into inner sensation, into something which happens to a *me*. To put this even more precisely: *the core self is the meaning of bodily processes*. It is their always-present inward dimension from which they cannot be detached. Consciousness, then, is not a representation or mapping of the body, as Damasio probably would say. It is a symbolization of living subjectivity by means of value, meaning and inwardness.

But what or who evaluates these experiences? Is there an authority responsible for sorting and ordering inner states? Is there some homunculus within the brain, an independent observer that screens and interprets all data? Certainly not. Organisms obviously lack such a control center, yet they are still able to synthesize and act upon a coherent perspective. Or is that really the point? That the absence of a control center is necessary to allow existential signification to unfold?

To grasp what is happening inside an organism, we need to understand the difference between representation and expression. In the old idea of the homunculus, this central authority screens a representation of what exists in the world. It browses through information and puts it all into some order (hence the tendency of scientists to liken thoughts, feeling and any cognition to information processing). Cognition is seen as the brain's way of representing objective realities of the world. A more persuasive account of how organisms make sense of the world relies upon embodied expression. Feeling is about an organism's experience of meaning and its manifestation through its own body. Instead of representing external information, the subject *expresses* what the nameless influences it is encountering mean in relation to how it is able to carry on, enfold and prosper. The "how" comes before the "what." The external environment that is there in an objective way is expressed through a setting which can be understood through its experienced relevance for the living subject. Experience is mediated through feeling, and feeling is focused in the core self.

The original experience that triggered this particular living gesture is physically expressed through a constellation of molecules, body fluids and neurons. These are matter and form at the same time. Their constellation symbolically encodes the original experience. This code is readable because it conveys an existential message on an emotional level. Like the paramecium under the cover slip, an organism that is doubled up in pain is a formation of matter that conveys the fateful personal meaning of existential suffering. It is only "stuff," and at the same time it is fraught with a sign of destiny. We can understand this destiny by virtue of our own emotions. That is the key: you have to be an organism yourself to be able to decode the gesture and know from the inside, without language, what writhing in pain looks like and means. You can only decode the gesture using your whole body and its entire history. On a basic level, therefore, way before any verbal explanations come into play, there is only a repertoire of gestures, which is more or less fit to convey the significance of a situation. For this reason, the core self is where poetic experience starts. And for this reason, the core self achieves a gestural recreation of meaningful situations through the poetic and expressive means of its own body. It means making the heart race in situations where time becomes a series of frenetic instants and letting the face grow pale when all the light has gone out of the world. We will now explore how organisms communicate with themselves using these poetic gestures of the core self.

EXISTENTIAL POETICS: TRUSTING ONE'S GUT

Living expression is something that any being immediately understands. In fact, it has no option to *not* understand. Its body reacts to the expressions of another body because it feels their meaning and seeks to understand the consequences for its own coherence as an organism. Emotions, not objectivity, count in biological world-making. What "really" triggers a reaction in an objective physical sense can remain totally unknown. Ultimately it is not important.

This shift of perspective is crucial for our understanding of life. In the classical, mainstream view that focuses on "independent," objective information out there, a representation must accurately convey useful information to an organism: a depiction of "what is" with all the relevant details. In organic reality, however, realism is not about copying or representing what is; it is about inventing an emotional gesture that can trigger a feeling or behavior which symbolizes the meaning of a stimulus or a relationship. In order to create a symbol — that which expresses something else — and convey a meaning, a gesture must be able to affect the behavior of a subject. It must have the same effect as that to which it refers, but it has to do this symbolically, by being totally different from that to which it refers. A bodily gesture, as the slipper animal showed, doubled up in destruction, conveys the corresponding feeling to an observer, although it is something entirely different than a feeling. A feeling is inward, and the bodily shape is exterior. Because living beings react emotionally, however, according to the meaning something has, the feeling is reproduced in them. For this reason, anything which conveys the feeling of "being doubled up" works. This can be done in any medium. It can be a sculpture, a picture, a musical note, a metaphorical description, the pitch of a voice. These modes of suggesting an existential feeling are the area of the arts, which from this point of view can be understood as attempts at re-enacting aliveness.

Acknowledging the symbolic reality of organisms enables us to reinterpret many things organisms do and to reassess why and how they do them. Consider innate behavior. The standard interpretation holds that to behave properly from birth in crucial situations, an organism, which can be as small as a fruit fly, must have every action that it needs to perform genetically encoded, together with an image of the related environmental stimuli. This model holds that an organism is basically a robot that orients itself using information stored in an interior map, which then has to be applied to the topography of the environment.

The neurosymbolic alternative to this schematic explanation is that a living being chooses which behavior to use because it is a subject with feeling, and therefore realizes to what degree its existential concerns are being satisfied by its behavior. Its actions are not based on stored information but on its own choices and adaptations to make it feel all right. Instinct does not work by clear rules. It works by trusting one's gut. It does not achieve its reliability through hard-wired information, but rather because it is grounded in freedom of choice. Only this approach provides the flexibility needed to respond effectively to an ever-changing world.

This symbolic theory of development and adaptation, which so far has been more or less neglected by ethologists, has at least three advantages: it presupposes a minimal burden of stored information that would otherwise make behavior inflexible; it shows how behavior based on symbolic world-making can be applied universally in any surrounding; and it proposes that behavior can be changed quickly by only a few genetic mutations. Remember the weak linkage between genes and embryonic development described in Chapter 3? This is an analogous situation, only it occurs at the level of behavior of a developing cell. Or think of language: the connection between the word "nerve" and the real bodily fiber is not the substance but the sense. We must keep this point in mind.

"Seeing with symbols" is how the human race has made itself a success story. The success of life in general is based on the fact that it is organized symbolically.

The core self is just as immaterial as the power that brings forth the cell's coherence and which maintains the plasma's identity, while matter passes through it. This power is not some force that Newton ignored and which we now have to introduce into our physical calculations. It is not the *vis vitalis* of the early 20th-century German developmental biologist Wilhelm Driesch, and it is not some external mind inspiring the physical brain, as philosopher Sir Karl Popper and Nobel Prize-winning neuroscientist John Eccles believed. Nor is this force an illusion, as many contemporary scientists still seem to think.

The core self is a physical reality, but so is the other, inward side of metabolism. Its force enables the ordering perspective, the standpoint of being self-concerned, an individual self. This standpoint is poetically symbolic. It does not *represent* the body as a control screen represents the operating conditions of a power plant; the standpoint translates its values into feeling. Translation is always poetic in nature. It requires finding something that is entirely different but nevertheless, by some logic, stands for another by being as relevant.

Feeling, however, is the only scale that can express what is relevant for a living being. "Relevant" refers to signification that conveys whether the life process is prospering, whether the poetic ecology of autopoietic self creation is succeeding. Feeling is the common language of all cells and all beings, the language of bodies and of poets. Only decisions taken in this language can have effects in the real world. And they can only work if the body retranslates them into the swelling of its muscles and the tension of its limbs. This is the decisive difference between the sphere of organisms and the world of machines. The latter is governed by circuits and feedback loops, which do not care about their fate. In the core self of an organism, however, everything is always at stake. Because the feeling of being alive mirrors the physical coherence of the self, the question is, literally, "To be or not to be." Its center is the desire for life. This difference lays the foundation for a revolution in biology.

SYMBOLS: THE LANGUAGE OF LIFE

In which medium does the desire for life articulate itself? First, we must admit that inwardness finds its place in this world through the core self, which is the existential symbolization of the bodily self. Inwardness is nothing supranatural. It does not go beyond matter. There is no divine spark from another sphere. The inwardness of organisms, and hence our own feeling of being a self, is a phenomenon unfolding in matter unrelated to a bodiless spirit. But where then is it seated? How can we grasp it? The solution to this riddle will help us to proceed deeper into the understanding of what consciousness is. Again, it requires us to invert our gaze. Our inwardness is nowhere in space, and yet it is everywhere.

Neurobiologists like Jaak Panksepp think that the symbolic sphere of our interior being expresses itself in one universal language, the lingua franca of the body. The Estonian researcher has an idea how the phrases of this language could be structured and, therefore, translated into our understanding. Panksepp again follows the basic principle that the constituents of inwardness cannot be reduced: the phrases of the lingua franca cannot be a text abstracted from matter. They must rather be a form in which matter appears so that it conveys a value that organisms immediately understand. This makes things nearly easy. The medium for feeling must be emotionally shaped matter. Feeling needs matter like fish need water. Without matter, the language of feeling could not appear — because feeling would not be there.

Such an idea again reminds us of how a work of art exercises its spell on us. A work of art is at once matter and meaning. It is the material arrangement of an emotional content by means of the senses. For brain researcher Panksepp, the lingua franca does exactly that. Its impulses produce expressive figures. The arousal states of neurons and brain regions do not represent or map emotions, and they do not code them digitally like a computer would be instructed to do. They do not represent, they embody. And this means that in some sense the neural states (which we can measure or visualize) figuratively *are* the feelings. They mimic shapes and echo rhythms. The core self does not represent; it symbolizes. It does not depict in a one-to-one mapping; it translates. And to enable the transmission of content, it tends to invent monstrous exaggerations and surreal distortions. The core self imagines through a pounding heartbeat, an eyelid's flutter, the torrents of spring blossoms, an early spring vocalization or the copper scales a beech scatters around its stem in autumn. It is as expressive as a demented genius working at his desk late at night.

The symbolic connection between consciousness, emotions and body, has far-reaching consequences — something that scholars like Damasio do not entirely fathom. To perceive mind as the symbolic correlate of the core self brings us closer to an answer to the question of how the strange ambient intelligence of mind happens. It makes it clear that mind is nothing different from body. It *is* body. But it is body in its meaning for the ongoing life process. The medium, therefore, really is nothing more than matter. But it is matter shaped in a way that has an irrefutable and immediate signification for a sentient being. The universal language is an existential meaning conveyed through expressive form. And the form leading to that meaning can be expressed by any number of expressive means. It can be conveyed by the rhythm of nerve impulses, by the moving body itself, by the species composition of an ecosystem, by a statue sculpted in three-dimensional space, by a sequence of sounds, by the arrangement of black and white spaces, by the choice of a word following another word.

For Panksepp, the communication of feelings works in just this manner. It is a kind of contagion through analogy. The arousal states of nerves instantly correspond to the contours of emotions. The firing of neurons activates the feeling. It is not coded digitally but shown analogically. The decoding device is one's own embodied existence. It translates the dynamics of an experience into a bodily form. In this way, the symbolic effect can be continuously transferred from one medium to the other — from the exterior world into the electrical vibrations of the neurons; from there into the tension of the limbs and the muscle tone; and then perhaps into a verbal utterance. Through all these transformations it never loses its symbolic content.

All these bodily shapes and gestures, movements of arms and legs, tension or relaxation of the skin, flutter of eyelids are not representations of emotions that have to rely on a binary key stored somewhere in the organism, presumably in the genes. Rather the body symbolizes emotions in the way a picture or a piece of music does, not as a strictly regularized system of signs but as allegory, as performance whose message poetically conveys what is enacted — the language of bodily gesture. The behavior that is triggered by neural fluctuations tracks the neural waves of tension and arousal, which symbolize the accompanying feeling. The form of neural impulses, therefore, is not a binary shorthand of the computer-brain. It is a poetic rendering of emotional content, a poem, not an algorithm. For that reason, feelings and consciousness cannot be explained as abstract micro-models of the organism or the self, as some neural researchers like Thomas Metzinger hold. How would this model function in the brain, after all? It does not have any device for virtual simulations. Feelings are rather imbued with the sense of self because they express the nature of its experiences.

Pankseep has observed deep similarities between emotions and corresponding bodily gestures, which he believes support the idea of the core self as an emotional intermediary between the inside and outside of an organism. Through the core self, inward feelings and meaningful outer appearances are connected. Outwardly, any emotion is inscribed in a nonverbal manner into a bodily gesture that is a perfect expressive counterpart to what is experienced inwardly, communicating richly without speech in a broad amplitude of ways, from hectic convulsion to calm flows of neural impulses. The inner state of an organism can mirror itself in its physical body because both are analogous. Both are shapes that a living subject is experiencing as the very texture of its self. There is fear that renders the body stiff and immobile. There is lust that manifests as rhythmic attention to the other. There is care when the body opens and tenderly receives the other. There is play in its contagious, bouncing lightness.

Panksepp's hypothesis is supported by further observations that the connection between emotions and gestures does not work in one direction only. The German neurobiologist Paula Niedenthal recently showed that a person who assumes a certain body posture (bold, timid, caring) will after some period of time feel the corresponding affects. It is more than a cliché. Someone who smiles a lot will tend to make herself happier. Every expression is at the same time an internal impression. And every experience inevitably expresses itself. There is no way it can be completely hidden, as it is how the existential situation of the whole organism inevitably manifests in its matter. Subjective affect is an objective description of an individual's existential situation. This principle is as basic as the law of gravity.

AN ESPERANTO OF EMOTIONS

When we follow Panksepp and his lingua franca, the world makes far more sense than we might have supposed. Under this view, we can assume that we human beings, living organisms from our first moment on, are endowed from the beginning with the capacities of making and receiving basic symbolic patterns of emotions. We are made to read the world of organic meanings. These basic patterns of feeling retain the same character from their birth as impulses in our metabolisms to their eventual expression in human language. What we experience as painful tension can be traced back to a fireworks of neurons that seem torn and flickering and that conversely mirrors the organism's reluctant shying away from the trigger of this tension. In a similar fashion, a work of art — say, a piece of music — will use similar analogies to express an equally painful tension. Its tonal sequences will also show aspects of tension, resolution and impending fractures. In an organism as in a piece of art, feeling resides in the coherence of the body, which is a material arrangement. The form of the work brings together a variety of tensions and orchestrates them into a whole, which can immediately be perceived as beautiful because it is alive with existential meaning.

The translation of existential meaning into meaningful expression is not linked to any particular senses. Painful tension (or joyful relaxation) can be experienced in any way, by any sensory input. Every organism unfailingly recognizes its existential value and reacts to it with the correspondingly uneasy (or gleeful) reaction in its own idiosyncratic way. All our sensory channels are apt to symbolize emotive dynamics. And they all have their own resonance frequencies, the peculiar gestalts that have a rich, necessary meaning for a certain kind of living being. Take colors, for instance. They are not neutral but rather linked to particular emotions. Think of a combination of colors, maybe green and pink. Most of us will we experience this juxtaposition as taut and contradictory (although I have to admit that as a male my color judgment is not very reliable). Or take the soothing effect of green, the cheerful influence of yellow or the cool feeling of blue. The effect that certain colors have on the body has been known for a long time.

Tellingly, neural researchers have found that the measured shape of electrical maps of brain states often retains the same form through diverse levels of magnitude. If a graph of electrical excitation is plotted in high resolution, its line is not smooth but jagged. The form of these micro-jags resembles the overall form of the graph, though in a smaller scale. This strongly supports the idea of a common language of emotions. One of these researchers is the French neurobiologist Michel Le Van Quyen, who works at the La Pitié-Salpétrière hospital in the research group Francisco Varela used to lead. Le Van Quyen calls this phenomenon "scale invariance," a term that means that the shape of a phenomenon is the same independent of how it is magnified. An example is the surface of a rock, which in a microscale shows the same ridges and summits as does a whole mountain range. Scale invariance means that there is a constancy of form independent of the scale applied.

To show this in his study of brain states, Le Van Quyen magnified graphic recordings of nerve action, consisting of superimposed graphs of electrical waves looking a bit like oceanic swell. He found that the brain pattern always shows the same overall gestalt, whether viewed at standard scale or in a highly magnified form showing minute single details. A tiny extract of a nerve impulse has essentially the same shape as a huge one. The line of the curves and cascades at a lesser magnification repeats itself at higher levels of neural waves.

These neural patterns are significant because they constitute a single, organic unit of nerve responses; they cannot be broken down into underlying, independent units. They are one holistic form and cannot be disassembled into separate biological atoms or "organic bits." "There is no difference between local and global," Le Van Quyen concludes.[4] The neural flow is not assembled from single parts, which piece-for-piece carry specific information. Their meaning comes from their overall shape. It is expressively solid. Even if you expose an organism to a microscopic clip of the original nerve impulse, it will convey the same message as did the entire impulse. The term for this dynamic is *holographic*. A tiny fragment carries the same message as the overall picture does.

Expression of this sort has no privileged locus. Its signification is everywhere. It is a gestalt. When the brain is activated in a certain way, the form of brain waves cascades forth in all areas. Its character, so to speak, infects the whole brain with its dynamics. The size of the signal that propagates this dynamic is not important. What matters is the efficiency by which the quality of the original trigger is symbolized. The symbol communicates via its gesture — its organic form — not by its details, nor by its dimensions. Nerve excitation patterns show constancy of form and are scale invariant. A neural symbol thus becomes a way of reproducing a gestalt, whose influence is not altered by its dose, size or concentration. Metrics fail. It's not the quantity that counts but the quality. Here, the medium really is the message. What makes a neural symbol efficient is how well it can translate the original gesture into another medium — into the motion of another body, facial gestures, a linguistic expression, a picture.

Astonishingly, what we discover here in the depth of the neural identity of an organism through a thorough analysis by the means of cognitive science, is utterly familiar to all of us. It is how we decode social reality every day as we make inferences about figurative language and body gestures. We know from our own lived experience that a sharp, tense intake of breath conveys the underlying feeling of the whole — in this case, pain. How else could it be possible that after very few seconds a listener is able to discern that a given piece of music is by Gustav Mahler, even though he only knows Mahler's compositions generally and has never listened to this particular piece of music? A tiny excerpt suffices to communicate the whole. Lovers are certainly accustomed to recognizing their beloved through the most minute gestures of the hand, inflection of voice or raised eyebrow.

It is striking that with these structures the brain shows the same behavior that complexity researcher Stuart Kauffman found in biochemical networks. After crossing a certain threshold, biological networks crystallize and form a self-organizing unity. Just as the protein elements of a cell can come together to form a new and higher level of organization, it appears that fluctuations of brain waves may function in the same way. By means of their complexity they autonomously form a whole that is capable of giving itself a structure and behaving coherently. A magnification of this structure at any scale shows self-similar patterns and thus everything that is needed in order to understand its meaning.

From this perspective, it sometimes seems that brain researchers studying the neural processes underlying consciousness are describing a gigantically magnified cell. And brain scientists indeed use some of the same metaphors Varela applied when he described the cellular life process in its simplest form as the "creation of an identity."[5] Rudrauf and Damasio also believe that the brain generally follows a pattern of "resistance against variability." It attempts to maintain its complex self-referential state against external fluctuations. "The brain dynamics create a state of tension and concern which can be described objectively and which can be the foundation of subjective experiences," Rudrauf and Damasio say.[6] Constructing an identity is an organism's strategy for self-preservation in the face of external stimuli that could destabilize the system. Such a concept is nothing less than the central idea of poetic ecology, in this case as applied to the brain.

The constancy of forms, therefore, is nothing unfamiliar to us. We do not just hear it in a composer's musical signature but everywhere in reality. It is all-pervasive in nature. It permeates our experience from the smallest scale on up. Think of an approaching thunderstorm and the anticipatory, excited rustle of the trees being bent by the wind and the roar of billowing waves on a lake or ocean. These are natural phenomena that poets since antiquity have used to understand and express our emotions. They could do so because such wild movements in their relationships to living beings have a common meaning, namely what this wildness can do to bodies that are exposed to its forces. The experience of these elemental powers evokes the image of destruction and at the same time the experience of movement and energetic enthusiasm. Air and water are so different from one another, and yet their dynamic movements can mean exactly the same thing. The swooshing motion of the elements brings to mind the hissing of our blood when it is moving quickly and pounding in our ears. The wind's frantic rush has the same symbolic form as the blood streaming through our body at a fast pace. The sensations of sounds, fluids and tactile experience are all linked through dramatic form, which is scale invariant across the different materials through which it expresses itself. This constancy of meaning moves through our diverse sensory channels as if they were one single ocean of coherent meaning. The wind whistling through trees has the same meaning as the sea's waves fleeing before a gale.

Both have the same swaying movement. In such gestures we perceive absolute forms. They are ur-phenomena of feeling in which content and contour cannot be distinguished. They are the feeling in its materiality, and they are immediately understood by all beings without explication or translation.

If Panksepp's analysis is correct, then form can freely wander through a whole range of media, from the physical to the psychological to the aural to the linguistic. A chain of analogies in different media links them all — the fluctuations of the cell's molecules with neuronal patterns, the tension of body postures, the significance of verbal expressions, the momentum of musical themes. Form is the medium. What we perceive as form always penetrates deep into the body, for the chain of analogies reaches deep down into the smallest vestibules of the flesh, which in turn express value as bodily lust or bodily suffering among many other subtle meanings.

THE POETRY OF THE SENSES

It is this constancy of forms that gives us a firm anchor to the living world. We do not need to define ourselves any longer only as the animal that can think, as the *animal rationale* of Greek philosophers and Enlightenment thinkers, who made such a misguided virtue of standing far above and far away from what they presumed to be mindless animality. Poetic ecology permits us to understand ourselves as the animal in which nature's feelings find a voice. We feel, as all nature does, but we can utter those feelings by translating them into words and gestures, which in turn make others feel. Man, therefore, is, above all, the *animal poeticum*, the poetic animal.

But this special quality that characterizes our particular way of being ourselves is not, as in the *animal rationale*, something that separates us from the remainder of creation. It rather puts us at its dead center. It puts us in the middle by highlighting that which is the central quality of all life in us and by articulating this central quality in our specific way, which is the conscious application of imagination. We, as all nature, express ourselves in, and understand the meaning of, the lingua franca of embodied meaning. We are part of this language. We belong to the world-inscape: inside and outside at once.

The body makes it possible for words to send shivers down our back, and for ideas to make us change our entire lives in the blink of an eye. For this reason it is necessary to stress that it is through the existential symbolics of the lingua franca that we are not simply creatures of the subjectivity exquisitely entwined with our body; we are also creatures of culture, that seemingly rational product of history of which man is so extraordinarily proud. Culture is our species-specific way of embodying meaning and communicating what life is about.

These thoughts have an important consequence: We might more easily understand how organisms function by means of art than by the explanations of mechanical science. In acting according to the meaning of an event, our behavior is not the cause of the event but a reaction of the body to a given encounter. A body adjusts its equilibrium and standpoint in reaction to an encounter. As we have seen in the preceding paragraphs, a body does not react directly to material causes but to their meaning — and also to meaning that has no material form, but is "just" a sign, an idea, a metaphor, a feeling. These are felt as negative if an organism perceives that they might disrupt or affect its self-coherence. Its reactions are inevitably gestural. In this sense, an organism's response to the world becomes equivalent to a work of art that expresses meaning.

Any contact with the surrounding world requires an act of imagination. Therefore, a sharp word can hurt just as much as a sharpened knife. Both push forward into the circle of inner balance that an organism strives to maintain and threaten its coherence and safety. It is not for nothing that we call a hurting remark "sharp." From life's viewpoint, painful words cut through a tissue of meaning. We can use the same words for both senses, physical and personal, not because of metaphorical conventions but because our feeling bodies convert both — the crushing utterance and the deadly knife — into the universal currency that is valid in every province of life, into its meaning for the ongoing existence of the living being. Our language responds to the ecology of these relationships. Its images do not follow the logic of spatiotemporal associations among objects. They obey the logic of bodies. This is why poets can explore our condition as subjects by means of language. The French poet Claude Vigée, for instance, can "hear green a young walnut tree."[7] He manages to capture the most delicate stirrings of first spring which we cannot see but nonetheless perceive as a faintly distributed sound.

The wisdom of languages has brought forth a plenitude of these images. They all translate a bodily experience into an existential value. These metaphors find their point of reference in an absolute realm of values, which arise in the feeling of organisms. Perhaps we should call these deep linguistic images ur-metaphors, following Goethe and his term ur-phenomenon. They are primordial phrases that are what they convey because they recreate an experience that forms the core of our subjectivity. Here the language game becomes serious. We understand expressions like an "icy" or "warm" gaze directly from the center of our body. The sun, for example, with its warm glow, has always been a symbol of life's abundance. A lively face "shines like the sun," and phrases like "you have a radiant future before you" retain traces of the original experiences. Indeed the warm afternoon sun shining on my body is really life-giving; it provides warmth and energy.

From this perspective we gain an existential ecology of poetry. Or rather, we can reconsider poetry as a first-person ecology. We can follow its metaphors with our body, even when it is not as directly expressive about physical reality and place as W. H. Auden's "New Year Letter" of 1940:

I see the nature of my kind
As a locality I love,
Those limestone moors that stretch from BROUGH
To HEXHAM and the ROMAN WALL
There is my symbol of us all.
Always my boy of wish returns
To those peat-stained deserted burns

134

That feed the WEAR and TYNE and TEES,
And, turning states to strata, sees
How basalt long oppressed broke out
In wild revolt at CAULDRON SNOUT,
The derelict lead-melting mill
Flued to its chimney up the hill,
That smokes no answer any more
But points, a landmark on BOLT'S LAW
The finger of all questions. There
In ROOKHOPE I was first aware
Of Self and Not-Self, Death and Dread:
There I dropped pebbles, listened, heard
The reservoir of darkness stirred.

HAPPINESS IS UP

The American cognitive scientists George Lakoff and Mark Johnson have been following the track of the symbolism of bodies for a quarter of a century. They have been able to prove that poetic expression is not the exception, reserved for over-sensitive souls and forgetful school teachers, but the norm. It is the benchmark of our rationality. Language does not obey a strict logic, as the majority of philosophers believed throughout the 20th century. Language is not abstract and rule-driven. On the contrary, the images we use in everyday language are full of unconscious references to the body and its irregularities and contingencies. They map the core self in a complex way. Because we do not look with the right perspective, however, most of the available insights remain hidden from view. Lakoff and Johnson, unlike their more order-obsessed colleagues, do not dismiss the imaginative and metaphorical dimensions of language as a sort of contamination. For them, this colorful and contradictory dimension is the most promising way to understand ourselves. Only through the opacity of the symbol can we become transparent to ourselves.

According to Lakoff and Johnson, all mental categories ultimately spring from bodily experiences. The two researchers see the body as the origin of all abstraction. Most contemporary scientists regard language as an arbitrary system of rules that allows us to express ourselves. Lakoff and Johnson, however, stress that most of our speech is suffused with the echoes of simple experiences. To communicate these experiences, we use images that describe the body in its existential circumstances and corresponding feelings. Language recreates the body in a symbolic dimension to convey what the living subject has felt on a nonverbal and preconscious level. The two cognitive researchers state: "Many of our most important mental concepts are either abstract or not very clear. Think of emotions, ideas, our concept of time. We must understand them with the aid of other concepts which we better understand — as, for example, our orientation in space."[8]

In this way happiness is always seen as something "up," as in the "pinnacle of happiness" or "seventh heaven." To speak about happiness in this manner reawakens the sensation of lightness that is familiar to anyone who has been happy, that is, all of us. Not surprisingly, Lakoff and Johnson observe many metaphorical connections in language: "Important is big, affection is warm, intimacy is closeness, bad stinks, help is support."[9] It seems as if our linguistic images arise as a kind of concordance to the life of our body — a dictionary of all the possible embodied emotional states. Lakoff and Johnson believe that we learn to understand most of these concepts in early childhood when sensual perceptions and motor performances are indistinguishable from the corresponding emotional experiences. The closeness of the father or any other functional caregiver *is* warm *and* provides satisfaction. Children experience both as a union. Only later, the authors hold, do we parse the experience and assign different aspects of it to different dimensions of reality, making it seemingly more rational. We forget that in the depth of experience, everything is an integrated whole and that we can only understand ourselves and our needs if we keep this connection in mind — and in the heart.

PLACEBOS: SYMBOLS OF HEALING

Humans can utter words that cut deeply, but we can also speak soothingly in nourishing phrases. Words and gestures can hurt, and they can heal. For this reason the discovery made by poetic ecology that signs and meanings create reality is not just a matter of creating nice poetry. It shows a power waiting to be discovered. If we were to embrace its implications with a little more courage, it could kindle a revolution in how we relate to the world. It opens up new vectors of inquiry that could help us solve problems inside and outside of "mere biology" that still confound researchers.

The fact that simple gestures can hurt or heal is related to a widely known medical enigma — the placebo effect. Doctors usually assume that a drug has a purely chemical influence on the body's own chemistry, a phenomenon that is not so different from adding a reagent to a substance in a test tube. They view the drug as a lever that triggers a certain switch in an apparatus. The maddening fact for researchers, however, is that some drugs have a beneficial effect even though the active drug substance is missing. The patient — and the doctor in a double blind trial — is only *told* that the active substance is present. So these "drugs" by themselves do not necessarily flip a switch. But still the mere belief that the substance is there seems to do so. This is called the placebo effect.

Generally speaking, it is still not clear what processes govern self-healing in an organism and what specific role drugs may play in this even though, in a causal-mechanical view of biochemistry, all of this should be easily explicable. But it is not. Apparently, it is hugely influential for a doctor to ritually declare that something is an effective drug; then, to some degree, it actually becomes one. The neutral substance acquires a kind of magical power. A doctor I once talked to told me that he sometimes injected his patients with distilled water, making them believe he gave them a painkiller when in reality they could not take more of it. This measure provided remarkable relief. Physicians are using the placebo effect in practice, but still they are mostly unable to explain what is actually happening in biological terms using the standard medical framework. They remain baffled — and use ad hoc accounts to justify what is going on, applying non-explanatory concepts like the "healer effect" or the "power of suggestion." The truth is that the logical framework of the natural sciences, which determines the range of credible possibilities in contemporary medicine, is unable to explain healing success that does not involve chemical substances or surgical interventions. The only available set of explanations for causes, then, lies in the world of psychic energies, which exists outside of the scientific worldview and is generally regarded as paranormal or supernatural. To evoke vague notions of autosuggestion is an evasion, too, because it simply shifts the enigma to another, mysterious level. If all healing in the end is only suggestion, what, then, is suggestion? What is the relationship between "imaginary" and "real" meaning?

The viewpoint of poetic ecology allows us to reintegrate the split consciousness that underlies these dilemmas. If we embrace a biology of living subjects, it is obvious that a placebo works just as any other drug — not as a causal lever pushing a button, but by conveying a meaning that is important for an ongoing life process. The drug and the healing gesture are not opposites; both are signs whose meaning triggers a process in the organism that leads to self-healing. The drug is not a lever that causes the organism to assume a different biological state, which then triggers a cascade of causal reactions. It is, rather, an image open to interpretation. Healing in this sense means the capacity to bring forth a sustainable autonomy of self again. It means to enable the body's ecology to self-create itself. It seems probable that healing is brought about not just by drugs but mostly by the organism itself. Drugs support this self-healing power. Recovery means to restore the undisturbed flow that can assemble an organism's matter from moment to moment into a new and stable identity.

This argument can be easily proven in experiments. A simple sweetener, for instance, can inhibit healing processes in animals if first given together with an immuno-suppressive drug for a certain period of time. When the immuno-suppresant is no longer administered, the sweetener alone still leads to a weakened immune system. The animal's metabolism learns to interpret the sweetness as destructive and responds by diminishing certain areas of its own healthy self-reproduction. This experiment reveals a second interesting result. To the degree to which sugar is interpreted as a sign of illness, it loses its attraction. The test animals who normally love sugar soon avoid it like a toxin — and it probably tasted toxic to them as well. The sweet savor of paradise can, through an organism's reinterpretation of meaning, turn into a flavor of doom. Sugar suddenly can be seen as its opposite. Even sweetness is subjective! What counts is how each organism comes to regard the degree of good or bad on its own individual scale of success and suffering.

Even though we share a basic existential understanding with all life, there remains a vast realm of unknowable depth within every life form. All beings are constructed differently, all individuals are distinct from one another, no body is like another. We all are similar, and we all are different. Jakob von Uexküll called the meanings by which organisms construct their worlds "magical Umwelten" — magical subjective worlds. Every organism, after all, has its own body-building plan, its particular sense organs and subjectivities. Uexküll used the adjective "magical" to highlight the noncausality of behavior, the fact that it is triggered by signs and not automatically determined by causes. The signs are magical because we still do not really understood all the signs and interactions that impel animals to construct their own magical worlds that enable them to survive, grow and multiply.

The poetic expression of nature is plain magic, not because it follows supranatural laws, but because it reveals that relationships, not substances, govern the biosphere, and these relations are not defined by space and time. Think of the spiraling passages of the larvae of tiny moths in September's leaves, which are familiar signs that autumn is approaching. Think of the routes that migrating birds or butterflies follow. Think of the sea turtle's ability to return to the natal beach of a tiny tropical island in the vastness of the sea to lay its eggs after it has traveled from one ocean to another and back. Think of the silent course of the spotted owl in the darkness, moving in ways that we cannot comprehend, in a space nearly without light. All these beings for Uexküll are figures that have magically encrypted their unknown, unknowable necessities. Their subjective experiences are not like wires that are used by outside forces to direct animal-marionettes, but rather like personal worlds which organisms learn to create and project on to the uncharted territory of reality. The unknowable can only be understood as metaphor. Subjective necessities are indeed "thoughts of nature."

Part Three:
I am Thou

In this section I will explore why the presence of nature is necessary for the healthy development of our own self. The relationship between ourselves and other beings is a deep mutual interpenetration on a material and symbolic level. We are not only part of nature, but nature is also part of us. To understand ourselves, we have to recognize ourselves in other living creatures. To be mirrored is a central element in the formation of human identity: a newborn can only learn to experience herself as a subject if she is seen by her caregiver and can identify with him. Nature as a whole is the prototype of this other, which lends our self identity. The more other living beings, which are not controlled by man, disappear, the more difficult it will become for us to achieve individual identity that goes beyond functional self-reference. We are threatened by an unexpected danger: the loss of the possibility to love.

CHAPTER 6:

The Question in the
Eyes of the Wolf

---

*I looked at her
and felt her watching
I became a strange being.*
— D. H. Lawrence, "A Doe at Evening"

The end came suddenly, although we had expected it for some time. It caught us in a dust cloud on the high pass road leading to Mehal Meda. We were about to overtake one of the roaring Borsani coaches working its way up the splintered mountain cliffs of the northern Ethiopian highlands as slowly as an asthmatic pedestrian. The dented bus in front of us, made in Italy in the 1960s, carried a roof rack loaded with piles of sacks, boxes and plastic bags. Their outlines were barely visible in the yellow cloud of sand and dirt that the old bus whirled up. The track curled narrowly uphill, one turn following the other, threaded into the creases of gravel fields interspersed with dusty, stony fields that yearned for rain.

We crept uphill with unnerving slowness and yet we had not been much quicker before getting stuck behind the Borsani. The housing of the rear differential had cracked shortly after our departure from Addis. Getachu, our driver, had to constantly stop the four-wheel-drive truck when we reached a new summit so he could check how bad the damage had become. At every stop, Getachu crawled under the car to see how far the fissure had proceeded through the bulge of cast metal. "The wheel might overtake us at any moment," he joked, when he was back at the steering wheel and started the engine again after one of those checks. I could not find out how bad it really was. After all, we were proceeding, if only quite laboriously. During one of Getachu's roadside mechanical check-ups, I wandered to the rim of the cliff and peered down into the valleys below. We had ascended to quite an altitude in the high, barren mountains. Below lay the fertile sphere we had left some hours ago. It was an otherworldly scene. Far, far down, tiny barley fields formed a freckled mosaic of silence. Crouched in between were small groups of round huts, whose thatched roofs looked like withered umbrellas.

Climbing upwards, we receded further and further from this archaic idyll. The closer we came towards the sky, the more barren, treeless and bleak the world became. Less and less could I see the paradisiacal farm landscape, which had slowly melted away down on the valley floor. In the higher reaches, the scenery had become a kind of limbo with its dried fields of barley burst open and glowing in the sun. Groups of gaunt men wandered seemingly without aim along the road. They wore threadbare green and blue suits and had slung turquoise or violet blankets over their heads to protect themselves against the sun. Sometimes we passed women in white dresses. They stayed behind on the dusty track like lost pearls on a string without an end.

Then, still a little higher up, after the needle in the spinning altimeter beside the rearview mirror had passed 3,000 meters, the landscape switched. Tufts of grass lined the road, bristled and grayish-green amongst the drifting whirls of dust stirred by the wheels of passing cars. We had left the area of intense agriculture and the mountains had gotten back their soft skin. Being stuck behind that Borsani became all the more fatiguing. Perhaps Getachu had grown impatient, knowing that only a short strip of road remained until we reached our destination. Approaching a long curve in the road, he stepped on the gas and accelerated into the dust cloud to pass the coach. When he saw the cows, he pulled the Landcruiser to the right and the rear axle broke.

A loud clank was followed by a cloud of dust whirling around us, hurling cliffs and blue-sky fragments at my field of vision. Before I could understand what was happening, we were already stuck halfway down the hill with the hood of the Landcruiser pointing towards the road. The hot metal crackled. Then silence. Startled, I turned my head. Behind us the hill gently sloped downwards. From its wiry grass thousands of high, slender lobelias emerged, flowering in burning orange like oversized torches. Still farther behind, on the opposite side of the valley, the mountains rose in an oceanic swell of blue jags and peaks. They seemed like a congealed sea of stone.

In the sudden stillness, in the presence of the mountains, the short moments after the crash seemed to stretch to an eternity. I had finally arrived. I was here, in the highland of Ethiopia, and I felt totally alone.

I was searching for the Ethiopian, or Abyssinian, wolf, *Canis simensis*. I had decided to write about this shy and relatively unknown carnivore when I discovered that only 400 wolves remained to roam the highlands of this old African nation. I wanted to see this strange animal before it vanished. I wanted to meet this relic of old grandeur and lost plenitude just once before it faded away. I felt like I was entering a stately manor that was slated for demolition, left to thoughtfully wander, for a first and last time, along rooms with floors covered in dust.

My trip was intended not only as a welcome but as a parting — an exemplary farewell. I could hardly visit all the species in similar situations — Grévy's zebra and the oryx antelope, the Siberian tiger and the snow leopard, the leatherback turtle, the blue whale and the great white shark, the orangutan in Borneo, the Iberian lynx in Spain, the corncrake on the German Elbe river and many, many more. We do not notice the desperate situation of these animals because they have not yet died out, have yet not vanished from the wild. But soon it will be as though someone were to quickly switch off a sequence of lights. More and more species have passed the threshold after which the number of individuals becomes so small that vanishing is only a question of time. *Click, click, click* — they will be gone, and there will be nothing we can do. Today we can already count the remaining few on one hand, like survivors of a monstrous shipwreck. The Abyssinian wolf belongs to those particular species, which, with their charisma, create the magical spell of children's books. A wolf!

A real wolf! It is one of those animals that little children use to trace the outlines of their reality and to stock their imaginary worlds. The wolf is one of those beings with whom humans learn to think. And the Ethiopian wolf was as good as gone.

I had coordinated my travel plans according to the suggestions of Stuart Williams. The wiry young researcher with the Center for Wildlife Conservation at Oxford led a project to protect the last remaining Ethiopian wolves in northern Ethiopia. I had visited him in his bungalow on the premises of the Lepra hospital in Addis Ababa, where I found him between piles of books and a computer screen in a setting that was almost European. Getachu took a while to find the house ensconced between narrow streets lined with low buildings and dense acacia trees. We had rolled along groups of mutilated patients squatting in the streets. The minutes seemed endless.

Every couple of months Stuart published a new, melancholic press bulletin about the near disappearance of wild, Ethiopian nature, about large-scale deforestation and famine, which seemed to raise no interest. "It is so depressing that I really crave to leave. Immediately, if I could", he said. "If I only could find someone who would do my work."

For Stuart, it was not by chance that in almost no other country was life expectancy so low, hunger so great and the darkness as bleak — or the landscape turned so utterly inside out. In Ethiopia the stony ridges of the mountains lay as barren as broken bones with a memory of a plentiful past. The people had devoured the velvet vegetative coating, leaving the soil covered with rocks. "In Ethiopia only one percent of original tree cover remains," Stuart said. His words became my mantra while I rode the rugged roads up into the highlands. Every time I looked at them the parched lands silently repeated what the scientist had explained to me. I understood that this country had passed beyond doomsday. The Ethiopian highlands looked like a post-apocalyptical desert. In the scab of its mountain slopes I could see what would happen if plants and animals left forever. It became clear to me, in a heartbreaking manner, to what degree the material and emotional well-being of people are related and how much both depend on nature.

When the car stopped moving, Getachu and I looked at each other. Then he quickly turned his face and spoke to his daughter Aster in the back seat, who had been allowed to tour with her father for the first time. During the whole ride she had sat silently in the back, humming odd melodies of Amharic radio smash hits in a low voice. Now tears rolled down her face, but she tried to smile. She stretched and answered her father in their strange, musical language. I got out of the car. The left rear wheel stood out from the wheelhouse in a strange split. Some 30 yards away people had gathered and watched us motionlessly.

I turned away, took some steps, and sat down on the hard grass. Silence. Sweeping silence enfolded me, in which nothing could be heard but the cold wind. Its motion rocked the flaming candles of the lobelias and sent rhythmic showers through the quivering grass. The air smelled of thyme, which grew in wiry patches on the ground. It felt as if the stocky flowers had dispersed throughout the air, like an aromatic essence dissolving in the bath. In the distance a gust drove a crow up the flower-covered slope, like foam on a wave.

Getachu had gone to the street and was talking to the men who had gathered at the site of our accident. Aster was leaning on the car and looking at her feet. The wind made her violet fleece sweater swell and ruffled the dark plaits of her hair. Some of the men started to unload our luggage. They took the tents, the dented aluminum boxes with the food staples, the folding chairs, the heavy bundles of half-gallon water bottles covered in plastic wrap. We had nearly arrived. The place that Stuart had proposed to set our camp on was only some 100 yards away, behind a couple of curves in the road, and then a tad farther into the rolling hills.

The men looked like beggars; they were starving. They were so skinny that I could easily discern the shinbone and fibula in their naked lower legs, legs no thicker than those of small girls. Some of them were barefoot; others wore plastic sandals that had been repaired many times with coarse thread. Their clothes, made from cheap uni-colored tissue, were ragged and had been mended many times. The bizarre green and blue blazers with matching shorts, that everyone seemed to wear, looked like worn-out patchwork. Through the holes in their clothes I could see their ribs protruding under the skin. I hauled the pack on my back. This small effort already made me breathe heavily. Here, at an altitude more than 1,000 yards higher than Addis, with every step I sensed the reduced oxygen content of the air. My ears buzzed and I felt the return of the slight trace of nausea that had ailed me in the first days after my arrival in the capital. Addis is nearly a mile and a half above sea-level. With the altitude sickness, a gloomy indifference returned, not unlike seasickness.

We trudged into the rustling grasslands. Soon the men began to assemble the plastic skeletons of our tents and pull the silvery skin over in well-practiced motions. One of them vanished with a yellow canister to draw water from a hidden spring, another returned with an armful of crooked wood for the fire. So we would contribute our share to the ongoing deforestation taking place here. While the men worked, I caught my breath for a long half-hour beside my backpack. Then I slowly sauntered into the hills. Aster remained at the camp side. Her father had stopped a lorry and now rode on its load area to Mehal Meda to organize our car's repair. I was now stranded, for an indefinite time, exactly where I had wanted to be. I felt dizzy. I could not leave anyway.

The wolf did me a favor, as if compensating me for the uneasiness caused by the rough ending of our journey. It appeared as soon as the campsite had vanished behind the hilltop. While walking I looked to the ground and my gaze followed the tiny, undulating corridors incised into the grass. They were the minute paths of countless rodents which populated the area en-masse. From time to time I saw a shadow scurry along the open tunnels. It paused, sat down with nervously quivering nose and then sprang up and hurried on. A whole crowd of mice and rats, common and bizarre forms, like the giant mole rat, lived in these alpine meadows feeding on the rich seeds of the grasses

— the Lady's Mantle and silvery Helichrysum flowers with dry stalks and leaves that compacted under my feet with a rustling noise. The scurrying crowd of small animals nourished by seeds and leaves of the dense vegetation offered rich prey for the wolf.

Sixty years ago the whole range of northern Ethiopia had been a bountifully laden table for *Canis simensis*. At that time, dense forests still covered most parts of the country at altitudes well above 10,000 feet. Still higher up, the soft folds of the Afro-alpine highlands and moors were shielded by a tender, protective skin of grass and aromatic plants. Today, only very few acres remain as they were. Such an area, still preserved, stretched out before me, in the Guassa-Menz region, 125 miles to the north of Addis. Other tiny patches remain in the Simien and Bale mountain national parks, which are generally in very bad shape, though.

When I looked up, the wolf was gazing straight into my eyes. I was petrified. I cannot describe what happened next as anything other than exchanging glances — we looked at each other. The predator's black eyes were set on me motionlessly. They stared at me from an abyss of otherness. But they did not convey anything unfamiliar. On the contrary, I saw a lot of myself in them — a lonely wanderer in the speechless mountains. How long did we stand there? A second? A minute? The animal had thick reddish fur that took a golden hue once its breast pointed towards me. The muzzle was long and elegant, like a greyhound's. At some point the carnivore veered away with a gracious turn of his body and progressed downhill in a straight line.

For a while, the wolf's path divided the gray herbs of the mountain prairie like a wave running through a leaden sea. The animal, already distant by a few hundred yards, vanished behind the next hill's crest. I still looked, long after the sea of blades and panicles the wolf had stirred had calmed. I did not have eyes for the rodents anymore. In these few moments the predator had created a blank space in my heart, a void that I had not recognized before. Where it was located, I missed myself.

I was left mesmerized by the wolf's gaze, left with an impression that was vastly different from all captive and domesticated animals I had ever encountered. It was as though it owned the right to estimate my value and weigh my personality, and perhaps to discover it too light. It was also a welcoming — but a welcoming to what? A salutation to the earliest past of my own species, *Homo sapiens*, which had slowly labored its way from nature to a place where it dared to oppose it head on? But this current about-face was preceded by an even longer period of common ground. For hundreds of thousands of years, man only understood himself as existing amidst all nature.

The extent of this past is nearly unfathomable. For 99 percent of the time that humans have dwelled on earth, their view of themselves was in the context of other animals. The motifs populating prehistoric rock art (and still existing ritual art done, for instance, by Aborigines in Australia) almost exclusively show other creatures. The first paleoanthropologists who tried to decipher the meaning in freshly discovered stone age caves, like Lascaux in France, have theorized for decades as to why our forebears mainly depicted other creatures (and rarely humans) on the dark walls receding deeply into the caves.

By now, human ecologists understand that early humans neither portrayed the animals with admiring reverence, nor did they magically invoke them as hunting prey. The reason for ritual art, it seems, is simpler than that, but on the other hand, more complicated. We need animals to think. By comparing the cultures of the last few remaining archaic societies in remote corners of primal forests or dry savannas, anthropologists have observed that natural symbols are used to express the human understanding of the world. The early humans used the ecological order of the living world to categorize reality in order to imagine and understand the cosmos. This ritual culture, which lasted for hundreds of thousands of years of human existence on Earth and which has been preserved in rock art, was a way in which the first humans comprehended their own nature. The rock drawings depict a symbolic ecology that anchored man amidst the rest of creation for hundreds of thousands of years. By learning to read relationships among animals and plants, and their connection to the local topography, the early cultures performed something similar to what happens when the symbolic mind emerges from bodily experience. They imagined their own lives by means of living images. They created using the forces of creation. They imagined a metaphorical system borrowed from reality, which, at the same time, was *about* their relation to reality.

This self-referential process of harnessing the already active symbolic powers inherent in ecological relations might even be the shortest definition of culture. Culture is a symbolic system interpenetrating the encompassing symbolic system, which is life. Culture stands in a relation to nature that is similar to the paradoxical identity between core self and body. Culture organizes ecological reality, but it does so in an explicitly symbolic fashion. Seen from this viewpoint, we can understand that ecological processes act as a proxy for culture: they are ordered in a way that keeps an interwoven web of relationships going. Accordingly, early humans perceived nature as a society with diverse members existing in a multitude of interrelationships. For this reason, the rules of archaic cultures were mostly conceived following what the human actors perceived as the principles governing nature. The understanding of these principles of nature, on the other hand, clearly was not an objective representation of reality but an act of the imagination. Nature with its expressive surplus met human nature with its imaginative power.

How nature was perceived by early humans followed these principles of the ecological imagination. Ritual pictures are, therefore, able to demonstrate how this poetic transformation of relationships works. We can observe that it is enabled as well as constrained by embodied symbolics. Water is crucial and good, for instance, and is thus represented in a sensually intense manner as "sparkle and force" in some aboriginal art. (see above, chapter 4). This living power is expressed according to the poetic freedom inherent in all life, and particularly in our species. The pictures on the cave walls illustrate

151

this encounter, which is an interpenetration of unconscious, embodied imagination by the ecosystem, and the imaginative answer to it given by one member of this ecosystem, *Homo sapiens*. They show reality exactly through showing it as an imaginative surplus, as an unfathomable source of creativity. They weave their mutual interpenetration into something becoming a living authority in its own right. The images express an abstract context by imaginative means, and therefore never lose contact with the central experience of life.

In Ethiopia this deep time had long passed. No other country has been Christianized so early, starting from the first century AD; nowhere else in Africa had humans exploited nature as quickly and in such an uninhibited way. In the 15th century, records already chronicled the catastrophic consequences of deforestation. Therefore, Ethiopia is not the ideal place to find people still living close to prehistorical ways of world-making. But there are still other cultures, where the Western disunion into *man* and *the others* was more recently imposed, as the Australian Aborigines already mentioned. Here we can still witness the connection between what is human and what is animate, which has been typical of human deep culture for at least 500,000 years. In some archaic civilizations, other creatures still lend a foothold to man's self-understanding and provide it with a form. Many examples still exist of how real beings become cultural metaphors because they show the same existential structures as human experience does. The members of these archaic cultures perceive the world through the bodies and habits of other beings. They use them as sense organs to experience themselves.

Let me explain this with an example. Some Australian Aboriginal tribal groups use to create ritualized "X-ray drawings" of animals. In such a drawing of, for instance, a kangaroo, the individual organs are depicted inside the bodily outline like they would be in an X-ray scan. The drawing usually has a host of different cultural meanings that emerge and overlap like the petals of a flower. The ritual painting of the kangaroo may superficially represent an anatomical design, but it can also be used as instruction for how a hunter should divide his prey. On a deeper symbolic level, the tribal elders might interpret particular organs as elements of the landscape called into being in the Dreamtime by the kangaroo ancestor through song. Every organ is thus enfolded into narratives recounting events, which had come about at specific places within the landscape. In an even more esoteric sense, these events are still taking place if the adequate connection with the ancestor is evoked. To paint the picture is to make the creative potential become real.[1]

All this functions through a complex interpenetration of spheres that were conventionally thought to be distinct. The human social rules are contained in the ecological rules that govern the life of the other animals. The landscape with its respective ecosystem provides identity to the humans connected to it because they are nourished by it and they experience its life-giving dimensions. The respective narratives — the Aboriginal songs — therefore, establish the rules for how members of a community can productively live together. Culture, therefore, does not mimic nature but follows its symbolic potential. According to this logic, in many Aboriginal societies, different social groups connect to distinct anatomical features of totemic animals and, through this, are linked to the corresponding features of the territory. In some tribes, for instance, different animals represent men and women. The differences between the sexes and their respective customs are also metaphors of ecological relationships between various classes of organisms — carnivores and prey, water lilies and fish, herbivores and plants, flowers and pollinators, different bird species that are active in the course of the year according to a mysterious pattern. From this we can intuit that a culture that understands itself in existential metaphors achieves a sort of holographic entanglement with reality, where cultural facets can connect to any natural detail, radically including organic structures into cultural symbols.

The deep experience that helped structure the thinking of the early conscious hominids is brought forth by the fact that all possibilities, constraints and needs in the human sphere already exist in the animate world. But they do not occur in the form of individual, human-like needs of particular animal characters. They subsist in a more abstract — or poetic — way. They are the laws that keep the ecological fabric together. The bee's dependence on the flower, the overabundance of young tadpoles in April, the lone mastery of the wolf commanding the composition of the fauna and consequently the flora, through its "ecology of fear" (the effect his sheer presence has on the behavior of prey species) — these are abstract rules of relating. And rules of relating also make up the human world. They do not describe the individual behavior of psychological characters but express existential modes of life-making-in-connection-to-others.

The strict laws of ecology thereby unfold into a cosmos of endless mutations. They become the condition by which ecological imagination becomes possible. They warrant that this imagination never ceases to proceed with fresh inventiveness. Nothing ever rests the same in the biosphere: fruits mature, herbivores are devoured, seeds sprout. One individual transforms into another according to a basic set of principles of birth and death entangling one another. In the end, the ways of our human world are nothing other than the processes at work in ecosystems of living beings. Personal life shows the same steady oscillations between expansion and contraction, between ecstasies and catastrophes. This is the scope of life. Nature, the totality of other beings' lifelines, is nothing different. Nature is aliveness. All forms of individual aliveness symbolically take part in it. Through this relationship of diversity-in-identity, the human world can be mirrored through innumerable natural processes. Humans are, through their bodies and through their metabolism, materially entangled with nature. Through their experience of aliveness, they are also symbolically and emotionally enmeshed with it.

It must again be stressed that this mirroring process is by no means trivial. Nature is neither the safe haven in which to take shelter from the ailments of civilization nor the perfect model for a healthier existence. Many idyllic views of nature, including a host of the ecological visions of the last decades, fall into the trap of construing their images from a human psychological point of view.

They anthropomorphize nature and forget that its power is not a superficial mirror of our emotions but a deep interpenetration of creative and imaginative forces, which can be destructive as well as benign. Our connection with other beings happens on a deeper level: it concerns basic principles of exchange — separation-in-unity, birth-through-death — which are neither exclusively human nor exclusively natural, and which are neither only natural nor only cultural, but which form the possibility of living form as such, the meanders of time on the face of eternity.

These deep principles cannot be verbalized, as this would untangle them from their interwovenness. They can only be experienced and passed on in a poetic process of imaginary fertilization. This unspeakable entanglement is what Theodor W. Adorno called the "non-identical," Theodore Roszak named "deep form" and Gregory Bateson termed "the pattern that connects." The truth of the world does not reveal itself in a linear-logical way but in symbolic-poetic means, and only this can bring a deep understanding and a practical social guideline. Poetic understanding means not always taking things as useful first-hand values but seeing the bigger picture, in which every loss is counterbalanced by a birth in another place and at another time. Take the proverbial horror of "nature red in tooth and claw," the fact that most species feed on others as per Tennyson's famous verse. In Victorian England this observation accorded nature the status of a tragic and deeply immoral limbo, waiting to be corrected by human providence and mechanical technology. However, many tribal cultures do not interpret the food chain as senseless murder but as a constant marriage. In many tribal thought-systems, eating another being is viewed as equivalent to sexual intercourse because it makes fertile with life. On an ecological level and observed without any Western human-moral bias, this tribal cultural idea trumps that of 19th-century England.

Food chains and other ecological dependencies thereby form archetypes of communication. This is an insight whose significance cannot be overestimated. In an ecosystem, death does not only

separate individuals but also inevitably links them together. All are
interdependent because each being consumes the other and can only
continue to exist in this merging of bodies where, from a metabolic
point of view, at some point one ceases to be one individual and trans-
mutes to another one. To feed and be fed are elements of a community
logic that shares one and the same matter — identity-in-multitude
— like the world itself. Death in early cultural symbol systems is often
not just conceived as a form of ruin but as creative fusion. In ecological food chains, archaic societies frequently spot a language that tells
of a community's intelligence and refers to an indissoluble entanglement of beings where mutual respect is is the first rule of growth
— the first step in fulfilling the three laws of desire.

Nature's patterns very clearly show that processes that are
seemingly contradictory can bring forth harmony on a deeper level.
Accurate communication is, therefore, always inspired by the multilayered meshwork formed by creatures and plants. Building new
relationships has to make every effort to not injure the delicate tissue
of the living. To do so, many archaic cultures have created complex
myths that call for sustainable behavior in the tangible world by
meeting the requirements of an imagined other — beings that are
not real in the sense we would understand but who play the role of
an overarching reference system, which embodies the principles necessary for meaningful ecological interactions. For example, ecological
and social rules can be mediated through the care of ancestors —
spirits that are neither physical nor only imagined. They are part of
this world, and they transcend it. The realm where the forebears of
human individuals supposedly dwell after death is not otherworldly
but present in the natural beings and landscape the original beings
entered into after the time of creation. Therefore, the creative forces
in archaic cultures are, from one vantage point, nature that has become immobile landscape, and from another, a still latent source of
power that is an active source of creation and with which humans
can connect. The Dreamtime of Australian Aborigines is not history; it is poetic space folded in current space-time where everything is

a sacred substance in reality. It is the raw mass of the ongoing formation of the world. At any moment it can exert its power, and with the right awareness we, too, can be in contact.

In the eyes of evolutionary biologist Edward O. Wilson and human ecologist Paul Shepard, we cannot disengage from the mental connection with animals.[2] Our mental interpenetration with the rest of life is our heritage from the way in which the genus *Homo* slowly squirmed free from its unconscious intermingling with all other beings. Humans have separated themselves from nature not by inventing entirely new concepts forming an abstract culture, which leaves no trace of their true origins, but by internalizing the rules of embodied existence and using them as a backdrop and inspirational ground for all abstraction. For the human species, growing out of nature has paradoxically meant being more strongly connected with it. The ability to form abstract concepts on which we often base our human difference to all other beings, in reality, is the heritage of nature in us. We have become independent by transforming the symbolic functioning of our bodies among other bodies into the more abstract relations of culture.

We can observe this process by looking at prehistoric art and artistic rituals still performed in tribal cultures. In these cases, we see that an artist does not arbitrarily invent the meaning in a ritual painting. The symbols he uses have a strong connection to necessity. On the deepest level they represent vital ecological relationships from which humans also cannot escape. The paintings are maps of life's principles and are, therefore, as real as nature itself. They are not mere symbols but a part of our flesh and blood. They are dimensions in ourselves that we can only grasp if we conceive them as elements of the living world and express them in a language of metaphors that can convey the forces of life in a manner our body understands. In this exchange, a human's relationship to animals is the same as that with its own body. Both yield experience and sense, are ways of connecting to the whole of creation, move autonomously and are uncontrollable in all their depths. An animal can show what a human is or, at best, what it can be.

This conceptual and emotional entanglement of human reality with the ecosystems it inhabits can explain tribal societies' reactions to losing their homes. Archaic civilizations understand society as the mirror and counterpart of nature the destruction of the latter also destroys the ties between humans and severs the connection of individual meaning and universal sense. Ecological destruction, therefore, does not entail just material hardship but also a metaphysical disaster. When people lose reference points, which can be visited and cared for, in a real landscape, they also become morally disoriented. This dilemma is broader than the deplorable fate of some exotic cultures that are already gone. It equally concerns Western civilization. If the archaic thought systems are correct about our conceptual connection to nature and culture as a creative interpretation of the inevitable interconnections with ecology, then any loss of nature menaces society as such.

NO CHILD WITHOUT ANIMALS

One particular clue makes the hypothesis extremely probable that animals and the "wild-thinking" upon which our embodied imagination is based are necessary for us.[3] This clue comes from our children.

Children are fascinated by animals. They crave to be nourished with animal images of all kinds. This is an undisputed fact in our culture. It actually seems that cuddly toys, plastic dinosaurs and dribble bibs embroidered with teddy bears stride ahead while their wild counterparts vanish. Nearly every kid's first picture book is crowded with animals. And this is a fact across all cultures. Even psychological tests used to assess a child's IQ apply animal symbols to evaluate perceptive capabilities. Indeed, many 12-month-old toddlers would enthusiastically throw themselves from their buggy just to touch a passing dog. Children are drawn to animals, like moths to a light. Why? Only a few authors have explored this. Amidst our elaborate concepts of early childhood development lingers a gigantic blank spot. Its presence is a consequence of the merciless underestimation and misjudgement of everything that cannot speak with words.

Nonetheless, developmental psychology has decoded many of the seemingly strange behaviors of the very young, discovering the usefulness of seemingly useless acts. Its message is that nearly all behavior has a certain organic logic, (even if it regularly drives parents to exasperation). Developmental psychologists theorize that children's behavior is driven by existential and embodied curiosity. Every subsequent action unblocks more and more intelligent behavior. Children are natural prodigies of pedagogical experiences. They unconsciously know what they need to understand in the world. Every crawl under the dining table becomes an expedition. Every game played becomes the seed for a new experience. Therefore, children can meet their own future anywhere. If this reasoning is accurate, then their obsession with animals must have a meaning. It must meet a deep need and satisfy necessary desires. Therefore, we, the parents, should provide our children's search patterns with the corresponding input. We should permit them as much contact with animals as possible.

But what do children find in animals? Human ecologist Paul Shepard argues that six-year-olds are still emotionally the offspring of prehistoric hunters. However, Shepard does not think this is due to mental hardwiring, which leads to an automatic perceptual need for the presence of nature, as sociobiologist E. O. Wilson suggests in his biophilia hypothesis. Instead, we exist as beings of flesh and blood and can only understand ourselves if we can borrow our thought-categories from an affiliation with the giant and unfathomably entangled web of other beings. They are like us, yet entirely different. Learning to relate to animals, therefore, means to become more deeply human and is comparable to learning to walk, which means growing a capacity for which the body-mind is made. Following animals with our eyes, our hands, our hearts, allows us to become ourselves. However, such a becoming cannot be explained as the mere execution of an inborn program. It is rather the longing of a body that cannot behave any other way because its logic stems from the way it is made, which is mirrored by other beings made of flesh and blood.

Language, often cited as proof of man's independence of biological reality, could not be possible without its intimate enmeshment with the body. Only the experience of having a body enables us to unfold the panoply of our categories of thought and feeling. George Lakoff's and Mark Johnson's insight, which I presented in the last section, is unequivocal: without reference to the body, our language could not develop. Our ways of thought do not spring forth from "pure reason" or from the arbitrary interplay of language elements but from the sentient logic that a body can die. Any linguistic abstraction is based on the most concrete of incarnate connections. To understand this body, which is us and which is us-in-connection-to-the world, we need the presence of and exchange with other bodies. We can only truly find ourselves within a web of other embodied beings as part of a rhizomatic multitude of subjects striving to create sense.

The most recent results of cognitive research show with scientific authority why the becoming brain needs this "other," another incarnate subject that consists of living matter, for its healthy cognitive and emotional development. It shows that we need an other to return our gaze in order to understand ourselves. We need the gleam full of impervious strangeness, which is able to enfold us like the wolf enfolded me with its dark-eyed stare, watching me, sizing me up and demanding nothing. We need the gaze sent from the aristocracy of being. We need it in order to be.

I turned back when the sun vanished behind the swell of the hills. The wind, barely warm before, suddenly became chilly and sharp, and dusk sank quickly, like a cloud of heavy steam. I trudged back on my own footprints in the pale carpet of dry herbs. In the gray, which sank more and more deeply around me, I seemed to walk in an endless snow landscape or through vast dumps of cold ashes.

As I write these words, lost in my memories of Ethiopia, I aimlessly look up and stop short at the sight of a fox passing outside my window. It paces along the scrubby rim of the flat, sandy Windmühlenberg hill with its scattered oak trees. It is just after 3:00

p.m., and I see the fox's reddish back. Its fur is soaked with the rain that poured down a minute ago. I notice the unmistakeable white tip of the tail. The fox is a lone traveler, determined and quick, gliding through the wet plants of my garden. It does not look up at me behind the window, behind the computer screen, although it certainly knows I am there. This animal is a cosmos in itself. It is a personality I can barely measure up to. In his behavior everything is necessary; it is Zen practice incarnate.

Before I moved to the rural district of Berlin where I wrote this book, before I regularly encountered the fox in my garden and in the small nature preserve behind it, I did not know the agile elegance this animal possessed. It slips under or bounds over fences with unexpected grace. The fox is slightly uncanny in its hybrid nature, which mixes qualities of dog and cat. Its character is ambiguous in many ways. It stalks by day but also by night; it is hunted by humans but not eaten, and it inhabits the fringe areas of abandoned gardens, cemeteries and the nightly streets of the metropolis. As I see it in its perfectly arrogant speed I understand that it only owns itself. Like no other organism in my urban environment it embodies the sublime independence of the carnivore, and with this the fox possesses something we know also exists in ourselves. Many of us secretly desire the fox's mysterious ways of having the world at his disposal. Therefore, his brief, enigmatic appearances vouch for what is most precious to us — our freedom.

When I arrived back at the camp, Getachu had returned. He had cooked spaghetti over the fire and prepared a pasta sauce in which I could taste the Italian influence in the old Abyssinia. I dined alone at a portable, plastic table that Getachu had assembled. The emaciated helpers in their tattered blazers and shabby blankets warmed themselves around the sparkling flames, their bodies building a somber wall, in which only their eyes gleamed.

Getachu had achieved nothing in Mehal Meda. The landline there was dead. He couldn't find a welding set with which to patch up the differential's steel enclosure. The next day he wanted to grab a ride to Debre Sina, the last big settlement we had passed. The town was basically an extended shanty settlement made from low barracks with rusty, corrugated iron roofs. There were few cars; the lanes overflowed with horse-drawn carts, which nearly broke down pulling vehicles on wobbling iron wheels without tires. Donkeys struggled under the high loads, set on their backs on wooden scaffolds. One had toppled under a burden of stones for street construction because playing children had hit him with a car tire they used as a toy. Only after several attempts had the animal managed to get up. I could see the blood on its back. "They load them like Isuzus," Getachu said.

We had spent the night before our trip to the wolf area in Debre Sina, in a small hotel with squat rooms grouped around a courtyard. Above the township's corrugated iron pile, hundreds of birds soared in the piercing wind, resembling a shower of huge, jagged flakes of ash. They were kites. As they looked for trash to browse for food, they enacted their aimless ballet in the sky, like rune signs of a nameless joy. When darkness had sunk, we walked over the sparsely lit square in front of the hotel to buy some water in one of the kiosks. A faded moon, this nightly star of Africa, poured its brownish light over the rubble covering the ground. Later in the room, under the shabby blanket, I was overwhelmed with the unexpected feeling of home.

The next morning in Debre Sina we made our last phone calls to Europe. We had to walk across the market to arrive at the telecommunications center. In the dust I saw the skins of freshly butchered sheep with their thick interwoven fur thrown down carelessly like old coats. On the flat roof of the telecom building, two speckled pigeons performed their courtship with unconscious elegance. A few windows of the faceless, square building were broken and boarded-up. The lady behind the counter handed me a form on woody paper printed in the cryptic fandangle of Amharic, which seemed like Biblical handwriting. On the dirty floor, the corpses of enormous brown butterflies were strewn, their wings adorned with vibrant, white eye-spots. They were like invitation flyers for an unknown feast that had been carelessly thrown away.

I sensed the stares of the men behind me and did not eat much. In front of my camping table, the pale Helichrysum mats stretched out like an undulating skin. The moon climbed above the hill, as big as a livid sun, and poured a mosaic of silvery light and velvet shadow over the plain. Its gleam lit the sky and made the layer of stratocumulus clouds fluoresce as though it were mother-of-pearl. The clouds skated over the night sky, icebergs on a black ocean.

Getachu had cooked too much food, and I barely touched my serving. As I stood up, the hungry men hurried to the pot with pasta. My legs were numb from sitting and hurt from the chill of the evening at this high altitude. On the way to my tent, I encountered Demessi, one of the hired guards watching over our camp. He was sitting on the floor, back resting against some stones, and he nodded. I did not see much more than the gleaming of his eyes and the faint light reflecting from the barrel of the old rifle that leaned against the boulders behind him.

At night, the altitude sickness came back. I could hardly sleep. I woke continuously with the feeling it was already morning, but the digits on my watch showed that only a few minutes had past. My ears rang, I was thirsty, and I felt sick. I opened the zipper of the tent. The sound ripped the black silence like a miniature explosion. I crawled into the open. The hard grass was dripping with dew that soaked my trousers immediately. I walked a bit through the crunching vegetation and stood alone among the mountains. The moon shone over the other ridge and turned the hills into a black and white negative image of a landscape. I sensed only the wind, this secret informer of silence. Nothing could be heard. It felt as though soundlessness had a shape that touched my skin, as if all the volume beneath the sky had filled with silence like with a thick liquid. I let it pour over me, flood into my ears and cover all the openings of my sense organs. It was an attack, a takeover by silence. I let it happen in awe. I was wrapped in resounding stillness. I looked up. The stars had become so numerous that they veiled the blackness of the sky. I was suddenly sure that, in this landscape, one would die with a sparkle of happiness in one's heart.

CHAPTER 7:

Learning to Think:
Mirroring the Other

Animals, you say. What do you mean? You mean everything
alive that you love because you do not understand.

— Elias Canetti

I n the morning Demessi's voice woke me up. He spoke some bits of
English and had come to my tent. "Emergency! Aspirin!" he said
in a dampened murmur in front of my entrance. My nose was cold.
I rose up in the rustling sleeping bag and pulled down the zip of the
front entrance, which made a ripping sound in the cold silence. Ice
crystals glistened on the silvery fabric of the tent. Demessi pointed
to Getachu's bivouac. In front of it was a plastic mineral water bottle,
a quarter of it filled with brownish urine. Aster was already up and
had kindled the fire.

I crawled out of my tent, walked over and flung open the nylon
web in front of Getachu's sleeping place. Inside, the driver was laying
on the ground, rolled into his blanket, with his daughter's spread
on top. He did not seem to see me. "What is going on?" I asked. He
slowly turned around his head. The blankets rustled. His face looked
very tired. He slowly said that he knew what it was. These joint pains,
the weakness, the high fever — malaria. He probably had not taken
enough Lariam the last time he had been hit by the disease. He had
led too many safaris into the tropical lowlands. A shiver made his
teeth chatter. He closed the eyes and slipped away somewhere I
could not follow.

Malaria. I sat in my tent and studied the information sheets for
the drugs I had brought with me and the medical brochure about
emergencies in tropical countries my doctor had handed me when I
got shots against rabies. I had a spare pack of Lariam with me. This
was pure chance. At the last minute I had chosen not to take this

drug as an anti-malaria prophylactic because there were so many stories among fellow journalists that it caused serious hallucinations. But I had kept it in my bag. Would the drug work if Getachu had never finished a thorough therapy because he could not afford the treatment? Was the parasite resistant by now? And what about the side effects? I remembered a colleague who had told stories about the frightening psychotic attacks he suffered after having taken the stuff. Like a really bad LSD trip, he had said.

No, Getachu whispered, he was fine with it. He supported it very well. But did he really suffer from malaria? Or from something else? We should see a doctor in the hospital in Mehal Meda. The place was the poorest shantytown in all Ethiopia, barren, no trees, rusty corrugated iron roofs. I had seen the clinic when we passed by in our four-wheel drive. It had no windows. "I won't go there," Getachu said. "I will rather die than go to the hospital." They would not be able to do any tests, they would have no medicine and they would probably kill him. I hesitated, and then agreed. Still it was a difficult decision. But in the end we did not even have a car to drive him to Mehal Meda. I pushed some of the huge pills through the aluminum foil. Now I did not have any more Lariam in case I got malaria, too.

Getachu's daughter made coffee. She showed no visible emotions. Later, during his agony, she lay in front of his tent and sang. The tunes sounded strange and sweet as if she was bringing the good spirits of the camp to voice. Getachu writhed in his sleeping bag as if he was dying.

I drank some of Aster's bitter coffee, ate some bread and set out walking. Demessi guided me. At every step, the buckle of the leather strap slightly clashed against the barrel of his gun. I wanted to walk over the high plateau until I reached the edge of the rift valley. Demassi basically knew no English. We hardly could talk to one another. I had shown him my goal on the map but I doubted he had understood me. I did not know if I would vanish in these mountains today without any trace. Or if my vanishing had already begun.

The morning sun quickly drove away the chill. When we had reached the crest of the hill we suddenly found ourselves in a group of baboons. The animals, big as large dogs, smacked and squelched behind a thorny thicket of the Abyssinian roses that speckled the slope in dispersed groups. As we moved on, the monkeys shrieked like startled chickens and galloped some 100 yards down the hill. Their silky fur glowed in the morning light. How clean they were, how spotless, how sparkling! When I was a kid I had learned in Hagenbeck's Zoo in Hamburg that there was nothing as filthy as baboons. And here their fur gleamed with limpidity. What a difference from the poor, starving people living in these mountains.

We went on through the crackling dry Helichrysum stalks. Silence weighed on the land and made every spit of the broken plants a singular event. A few yards away a swift's wings sharply cut through the air as if it was a tensely stretched textile web. Light patches fled over the speckled slopes and left behind a hint of warmth. The grass shone in a dull yellow. The stones were gray and white under their burden of lichens. Moss streamed in dull beards from the single stems of tree heather, tousled by the wind like the torn clothes of the shepherds. Close to the rim of the stone cliffs we met a single man who slowly climbed downhill. The wanderer scrambled down into the valley. I could see the irregular mosaic of the small yellow barley plots stretching more than 1,000 yards below us. We halted him. Demessi tried to translate. The man was on his way down to look for work in the harvest. His children were starving. As we were talking to him he again and again pulled together the riddled blanket over his chest. The outline of his ribs protruded through his skin. His legs were as thin as those of a young girl. I gave him a bank note, which I immediately thought was way too small. The man fell on his knees in front of me, clenched my ankles and pressed his forehead into the dust. I had to make an effort to make him stand up again. Then he adjusted his cloak, pulled a fold of it over his head, and left. The mule track in the steep mountain flank that he took was barely visible. Between its spirals Lobelias raised their tufts, faintly reminding me of pineapple blossoms. Slowly the traveler's steps trickled away in the fissured slope.

Demessi saw the wolf first, only because the animal started to whimper and whine. His body, maybe a few hundred yards away, was difficult to spot. He lay in the dried bed of a creek. His reddish fur was barely distinguishable from the color of the earth. The wolf howled, head raised, upper body resting graciously on slender forelegs, muzzle opened to the sky. The sounds surprised me. How playful, how dog-like they were! The strange impression was intensified by the long distance and the dampening effect of the dried mountain floor, and so they seemed unrelated to the movements of the animal's snout. I stared at the wolf. We had been squatting here for half an hour, and all the time I had been gazing at the rolling grasslands through my binoculars. The animal had stayed there all that time without a move. After a short while Demessi's face again took on the absent expression it had shown last evening when I encountered him during his night watch. The animal interrupted his baying, looked at us from below and turned away his head when we did not move. Then he jumped on his four legs and slowly paced away.

When we were just about to leave, we heard another yapping. This time it came from the opposite side of the valley. I searched with the binoculars and spotted a second wolf lying there in the grass. Then the previous one appeared again. They greeted each other like young dogs do. They wagged their tails and scuffled a little like puppies. The bright coloring on their chests flashed in the sun. The one who had rested in the plant cover was bigger. He stretched, stiffened his hind legs and yawned. He sniffed the air around and then made a playful attack towards the other who receded in a dancing manner.

Half an hour of light-hearted play followed, a puppy-style performance that filled me with joy. What naturalness in all of their gestures, what a self-evident display of presence, enacting their just being here, their just being alive. With all their movements the two wolves showed that every stretch of the grassland here was their home. Their empire.

Their playground. They could roll in its rustling gorgeousness whenever they liked to. They were the lords of these mountains, and they
were their children. How much enjoyment, how much enthusiasm
these beautiful animals showed! The wolves barked in a ballet of living
joy. They were the rulers of the hills, and they were the hills; they were
the hill's manner of enjoying themselves. My own joy was also part of
it. I also was invited to be part of the mountains for a short while.

WHAT BABIES THINK

"Joy of life?" Shouldn't we use such a term only with the greatest care
in relation to animals? Should we perhaps not adopt it at all? Or is
the opposite true? That only the use of this category lends us the key
to truly understand the behavior of animals, and above all, its meaning to us? Should we deliberately extend the sphere of experiences so
far reserved only for us and include them?

How greatly scientists' opinions vary on the matter as to which
cognitive abilities a species can have is best shown in the case of
our own kind. Only a few decades ago developmental psychologists treated newborns, infants and toddlers as not very much
different from animals. The valid doctrine read: as long as the
capacity for language has not yet developed, a child is not able to
have any image of the world nor of itself. It does not even feel, so
the teachings went, at least not consciously. It experiences at the
most something like unshaped affects.

Psychologists thought that self-consciousness, the crown jewel
of human self-enthronement, would only set in when a child could
consciously say "I," usually between one and a half and two years.
Without possessing words and the according mental representations
an infant, so went the dogma, could not distinguish between herself
and the surroundings. Her ego, governed by mighty but unconscious
instincts, was thought to extend to the whole world. In this phase of
"primary narcissism," which still today forms the ground zero of how
psychologists think personal identity unfolds, a baby is believed to
experience her mother as an extension of herself. (It is interesting

and revelatory, by the way, that the father is never conceptualized in these terms.) In this perspective, the mother is experienced by the infant as a gigantic attachment to her own body, which gives pleasure to the baby who is suckled by it, is enfolded in its embrace, is held by it. Without being in possession of words and the according mental imagination, so the classical position goes, an infant is not able to distinguish between herself and the environment. Psychoanalysis in particular still refers to this model today, calling it "primary symbiosis." In the psychoanalytical logic, every human inevitably will suffer a kind of original trauma because this symbiosis must necessarily rupture during development. Distortion and neurosis are inescapable. From the psychoanalytical standpoint the whole society necessarily is a prison from which no escape is possible. Damage is built into the human biology.

This story of the symbiotic unity of the child and her mother has trickled so deeply into our everyday psychology that nobody asks if it is true. Are we all necessarily separated from one another and from the world by a traumatic act, by an original loss impossible to ever repair? Or are we separate-but-in-communion from the beginning? Is loss necessary, but not necessarily traumatic, because it is the other side of connection? Have not Freud and many of his followers gotten something terribly wrong when they extrapolated the damage happening in the toxic upbringing, which was the norm in the bourgeois society of their day, to a standard model of how human identity unfolds? Doctor Freud, it must be remembered, started out as a rather deterministic neurologist and never left that law-guided approach in what became his picture of the psyche. But we might explain identity and how identity relates to the other, and how it comes about through the other in a different way, a way that centers on emotional experience and does not give up the primordial idea of poetic ecology, that life always is a problem and a solution at the same time.

For a long time a line in cultural anthropology has explored the idea that humans are "naked" animals, meaning that we lack something important from a biological standpoint. But in its dualism this approach misses the mark. All organisms are defective, so to speak, because life as such is centered on a fundamental deficiency: the matter and energy needed for the next moment are not there; they must be actively searched and built into the body. As I have tried to explain in the previous chapters, this lack of assured existence is precisely what starts the imaginative engine of the living.

There are only different ways of enacting the missing balance, missing in all life, because the void it leaves is the very motor for the desire of subjective self-creation in all organisms to unfold. I have thought about this fundamental constellation in my 2014 book "Aliveness" (*Lebendigkeit*), which is not yet translated into English.[1] From the position of a poetic ecology identity formation is a problem, because it is always a necessary mediation between self and other. But solving this problem at the same time generates the creative surplus all life builds upon on a material and symbolic level and which in its species-specific way is the fuel for human culture.

The loss so many humans sense so acutely, therefore, is not due to some fundamental existential distortion in human life alone (which could be remedied by "nature's harmony"), but happens only if the first and most important intimate relationship to the caring parent has gone wrong. In a primary bond that becomes toxic, the primary caretaker after birth is not able to live up to the unconscious biological relationship that the maternal body had provided for the baby. In this bodily interpenetration, a self is in relation, and in accord, with another self and both realize themselves through the other. The gestation period from this point of view is a good example for a productive ecological relationship. The parent's care of the newborn infant should transform the structure of this relationship, which now is no longer managed by the mother's physiology. The caretakers need to continue the nourishing and bonding that has taken place in utero through an active praxis of relating. This is the way culture takes over the co-creativity of organic affiliations. As we know, it all too often fails in this. But that does not mean that humans are flawed from the beginning, although there is a strong current in

Western thought that is biased in this direction, from the Christian dogma of original sin to Desmond Morris's idea of us being "naked," albeit deprived, "apes." Humans, so the chorus repeats, are lacking in something natural, so they need something artificial to compensate for this lack, leading them to occasionally overcompensate.

This attitude is another example of the widespread search for the special difference between humans and the rest of beings. But instead of being transfixed with separateness, we might understand oiur lives better by recreating in a human symbolic fashion something that is there from the beginning although it is easily lost. This perspective would entail that culture should rather be viewed as self-expression, trauma work and necessary symbolic-biosemiotic self-repair. Its role is not to repress trauma through an artificially created rational and presumably less flawed universe, as in the models underlying the idea of the naked ape and of primary symbiotic disaster as well as in the lion's share of self-definitions Western man has worked out. The idea of an inborn defectiveness which needs to be countered rationally ultimately posits all nature as a cruel "Leviathan," as Enlightenment philosopher Thomas Hobbes named it, a monster making us undergo separation after separation, that can only be fled or extinguished.

This is exactly the cultural heritage of the West. We can suppose, however, that such an approach for centuries has reinforced our separation from other life forms and from a huge part within ourselves, and that rather this constitutes our modern tragedy. The bold assumptions that have determined for many years how modern man understood himself would have looked differently if scientists had not divided their world into two sections. Still today in these theories the rational agents who are able to use language (ourselves from an age above three) are opposed to the impotent reflex (our and all other beings' bodies), which are condemned to solely unconscious reactions. This segregation of reality caused some startling side effects. Infants for instance underwent surgery without anaesthetics as recently as the 1960s because the doctors firmly believed that they were unconscious and hence could not be said to feel any pain.

While psychoanalysts still fight the traumas of broken primal symbiosis in their patients, cognitive researchers have made a series of fascinating discoveries, which point to a different picture, a view in which a human is an individual from birth on. There is no primary symbiosis, although no individual identity is conceivable without the life-giving presence of the other. But the relationship between self and other is more complex than earlier psychologists thought. It is less a symbiosis with an inflated maternal self, leading to a painful rupture and a lifelong longing, than a continuous, highly problematic but also highly creative, reciprocal creation of two autonomous yet dependent subjects.

The North American scientists Andrew Meltzoff and Keith Moore have done a lot to substantiate the idea of this primary autonomy.[2] They found in the mid 1990s that newborns are able to imitate human facial expressions some minutes after birth — and not only those of their mother's faces. Babies still wet with amniotic fluid gleefully stuck out their tongues when researchers showed them their own. As answer to other grimaces infants grinned or wrinkled their noses with a degree of compliance that made it seem that after the labors of being born exactly this activity of facial gymnastics gave them the greatest pleasure. Aspects of this imitation behavior, Meltzoff and Moore claim, are the basic glue linking the baby and her bonding person. They play out all the time while the caregiver and the infant interact. Through this gestural reciprocity, both start to tune into one another on a bodily but also on an emotional level. The nature of this mirroring is crucial. If the caregiver is routinely stressed, staring angrily, looking away or being anxious, the baby's emerging identity is seriously damaged.

Their observations not only surprised the researchers but also posed a dilemma difficult to solve. They asked themselves: How is the infant able to control his imitative behavior? He has never seen himself in a mirror. How can he know that he is the same thing as he sees in front of him, a material body with a face, that circular device with three holes and a hump in the middle, which so funnily distorts in front of his eyes? How can a baby understand that what he feels as inside, from a personal perspective of concern, also has an outside, a face of his own, which is of the same quality as the ones he is fascinatedly staring at right now? Meltzoff and Moore quickly ruled out the possibility that the perfect imitations the newborns delivered were only facial reflexes. The babies' behavior was far more flexible than could ever have been achieved by a mere inborn behavior like that which young chicks manifest by opening their beaks when the adult bird lands on the rim of their nest. The grimaces the researchers recorded in their experiments where too far from being stereotypes. The babies obviously enjoyed playing around and imitating anything. But neither could the infants have learned their behavior. They had been in the world only for some minutes. What they did originated in them — not in the form of an inborn reaction but rather as autonomous activity. When had they acquired it? During birth, instantly? That was improbable. In the womb, then? But in which developmental stage? At seven months? At five? At six weeks when the first human limbs bud? Or before that phase, when they had been floating through the dimly red space of the womb as a strange starship made from a small number of cells?

The riddle's solution probably is simpler and at the same time more enigmatic than psychology and biology have admitted for a long time. It is simply this: children *know* that the sensations they have inwardly, outwardly are a body, which is not at all different from that in front of them. This means that from the beginning, newborns experience their mother as another individual. And they know that her body — or that of a cognitive scientist, for that matter — is endowed with the same inwardness, which lets this individual feel from the inside what you can read on his outside. Newborns may not know everything, but they know one thing with a clarity which we can only envy — they are living beings whose feelings have a visible and sensible outside that can be felt with the fingers and seen with the eyes and smelled with the nose in the darkness. Or, put even more generally, they know that they are matter, which is capable of feelings. They know that they are alive and how it is to be alive.

If we cannot decide when the infant's ability to imitate sets in, we have to accept that it might be one of the earliest characters of a human being. In the last consequence this means that children simply "grok" what it means to be a subject, and they capture it in a totally nondualistic way by their inwardness and by their sensitive and expressive outside. An infant, therefore, does not play back a ready-made genetic program in which behavior patterns are moulded as fixed psychological templates. It knows how it is to be a specifically shaped piece of the world by virtue of being it.

Many researchers in other fields, however, do not really accept these results from developmental psychology. Therefore, they still cling to the myth of primary symbiosis with its connotation of infantile dumbness, or they bluntly adhere to the idea that knowledge of one's own identity and that of others from our double perspective of body and inward self is simply a hardwired behavioral program. The cognition researcher Jerry Fodor for instance states with the chutzpah of a successful computer nerd: "Here is what I would have done if I had been faced with this problem in designing *Homo sapiens*. I would have made a knowledge of commonsense *Homo sapiens* psychology *innate*, that way nobody would have to spend time learning it ... The empirical evidence that God did it the way I would have isn't, in fact, unimpressive."[3]

But the clues that speak against this are considerable. Inborn knowledge of the world is not the path chosen by natural history or "god" to enable organisms to cope. It is far too bulky and inflexible. Therefore, only the initial conditions from which the body structure develops are inherited. In the beginning of an infant's life there is only experience, the experience of being here, full of awe for that which simply happens — a living being's basic experience of being in the world. An infant unfailingly knows what it is itself — not because it already is provided with the modules of basic human psychology, but because the features that give rise to psychological experiences and which in the end make any relating possible are based on the experience every embodied subject has. The clue to folk psychology is not that it is ingrained but that it follows from the basic embodied experience with a sort of existential logics.

Because an infant knows from its experience of itself that it is inwardness with an outside, it can understand the outside of others' inwardness. This leads to an entanglement of its own budding identity and that of the other. The child understands what the other is because it is self, and it understands itself to the degree that the other returns this exchange of experiences. For this, the only necessary condition is that the infant is seen by his caretaker, that he is perceived as his self unfolds through its relating to others. By the enmeshment of inwardness and the outside, which is provided by all embodied experience, the infant understands the other and understands himself through the body of another. He can translate his own experience from the expression on another's face. And he can retranslate the emotional expression of another subject into its own. Because he feels what it is to be a subject, he is able to understand other subjects.

All this does not require any innate program. It only requires having a body which is vulnerable and which strives for continued existence and broader unfolding. It does not require learning something categorically new. It depends, however, on being allowed to be what one already is in order to become oneself in a more intense, more explicit way. Being a subject is learned by already being a subject. And it can be unlearned by being denied the permission to be a subject, by the exchange of identities not being reciprocated.

If somebody winks with his eyes, the infant understands from its inside-out that this is a movement, which is connected to a certain feeling — to the sensation of closing and opening the eye again. Researchers encouraged babies not only to push forward their tongues but also to cock their mouth, to pucker their lips or to slightly open them, all by imitation in the first hours after birth. But what tells an infant which gesture belongs to what inner sensation? That what he sees as mouth is connected to his own inner feeling for mouth? And what he observes as eyelids is related to the feeling for

176

closing these delicate shutters? What is this inner unity, if this is not a behavioral pattern, not a genetically fixed instinct, not a conscious deliberation? How does it feel to be a baby? In the center of cognitive science today we find the living experience of a being that most researchers a few decades ago thought to be not much more awake than a toad. What newborn humans show us brings us closer to the innermost mystery, closer to the question: What is a subject in its inner core?

ONE SENSE FOR ALL SENSES

Meltzoff explains an infant's imitation behavior by assuming that newborns — and so all humans and possibly all beings — sense their embodied existence in a "supramodal" or "synesthetic" manner. This brings us back full circle to Jaak Panksepp's argument: the core self, in which for him all existential experiences concentrated, is also supramodal. It comes before the separation of the diverse senses.

The terms supramodal or synesthetic both mean that the same sensory experience is made in different sensory channels and therefore, can be present in different perceptive modes. There are some people (like me, like my daughter Emma) who are constantly aware of this. Psychologists call us synesthetes. A synesthete more or less strongly experiences a color also as a texture, a sound, a taste and a smell, and vice versa. A smell, for instance, can be smooth or prickly, and always carries a certain hue or tone. Only one percent of the population carries this strange ability to see through the variety of senses.[4] In reality, this skill might signal a weakness in shielding off the unity underlying all perception in everyone. The fact that we all more or less agree on what warm and cold colors are points to the fact that all adults have access to the phenomenon. For a true synesthete, however, even abstract entities carry sensual halos. Numbers have hues and tastes and names are colored. Sometimes I ask my daughter, who apparently has inherited this perceptive sensitivity, if "freedom" is gray or blue, or how a Monday (orange-red, ouch) or a Wednesday (light blue, for sure) is colored. I once met a charming

British fine arts student called Sophie. As we quickly discovered, she also was a synesthete. I liked her name because it has a beautiful purple color to me. "That's what my mother feels and why she picked it," Sophie said. "But to me it is screaming yellow. Totally ugly. I hate it."

Like Jaak Panksepp, Meltzoff and his colleagues think that all sensory experiences somehow speak the same language *before* they are mapped according to the different sensory channels of seeing, hearing, tasting and smelling. We first perceive our own gestures and movements from the inside through our overall feeling of their existential value. Our core self thus is present in a synesthetic manner. The tone of the voice, the tonus of the skin, the movements of the limbs — these patterns of being can be from distinct fields but still carry the same value according to the subjective core self of the infant. Supramodality truly means to directly experience the existential value of any bodily gesture. The infant's core self is the translation device, which mediates between inside and outside and between self and other.

Accordingly for Meltzoff, not only can the different inputs from the infant's own eyes, ears, skin and palate or the stretching and contracting of the baby's body carry identical emotional values. The deep existential identity of the senses is even more radical. It includes not only experiences that the baby's own body is subject to, but also those of other individuals. This capacity does not need to be learned but comes from having a meaningful body in the first place. The newborn can read the same emotional valences in another person's gestures as he experiences through his own feeling. In one respect, the other and me are experienced as identical; in another, they are clearly distinct. Otherwise the imitational genius of a baby cannot be explained. It knows other, because other in some way is inside the self of the infant. To be able to perceive this way nothing needs to be learned. Empathy is generated as a resonance frequency of the self's own sensitive body. For Meltzoff, therefore, it is clear that the different sensory channels and the areas of self and other are not put together in hindsight. They are identical from the beginning. "We can metaphorically say that the perception of others and of self speak the same language from birth on; there is no need to 'associate' them later," Meltzoff thinks.[5]

Again we encounter here the lingua franca of embodied existence. It not only bridges the separate dimensions in which a subject experiences itself, mapping them in the currency of existential value, which is the feeling of being alive. The organic lingua franca is also the binding connection between different subjects. It expresses the experience of a subject's own meaningful inner experience by means of his body. By the same manner in which it can be expressive of a felt meaning, it also grants access to another's subject's inner states. These in the end boil down to the signification of felt existential impacts, which have values but not yet forms. They are pure poetic space. To the common idiom of subjective existence, therefore, inside or outside, self or other, gestures and words and material impacts are the same. All these neither need to have the same form, nor do they need to consist of the same material, but they still have an identical effect by the way they make the ongoing existence more or less possible. In this respect the idiom of the lingua franca always remains identical from the very beginning on — in the foetus, in the infant, in the adult and so in every living subject.

But what semantics does this language follow? What language is more general than each code of the separate senses, encompassing a wider scope than solely that one that is generated by your needs or my own? What language is eloquent enough to express inside and outside, feeling and body at the same time? It is the language not of things but of felt values. Feeling comes before form. The baby does not perceive the face he sees in front of him in the beginning as an object at all. At first sight the face is not an anatomical detail of a physical body but rather pure meaning. It is an impact on the core self. It is the consequence that it will engender for the self. The face embodies a felt value in its absoluteness, but only through being a body is it able to manifest this absoluteness. The gesture does not *show* a feeling. It enters the inner perspective of an infant only because, inside of him, it *is* a feeling. Each bodily gesture discloses itself to the infant only because it is experienced in the form of her own emotion. It is perceived only insofar as it becomes a feeling in the inward sphere of the baby. After considering this, we need no longer be astonished that we need only to look at a child gloomily to make it burst into tears.

INTERBEING: WE FEEL INSIDE OTHERS

Developmental psychologists understand that small children's talent for imitation is not mere play, but an activity that is crucial for their survival and healthy development. Meltzoff and Moore assume that the becoming self is imperatively dependent on those characters in the environment that are like itself. Only by relating to these characters — namely, subjects — is it able to develop its diverse sensory channels and its own identity. The infant not only needs an other as a scaffold for his own thinking and his own identity, but he must be in a constant mirroring relation in order to develop the diverse senses with their different possibilities of experience, their different ways of relating to the world. The case of children who are born blind or deaf shows that the senses have to be learned and are not hardwired in the brain. These children often grow up behaviorally normal, although their handicap makes it difficult for them to participate in the interplay of imitation and communication. For them, elastic development is possible because human sense functions easily can be taken over by one another. The brain stores the meanings of these substitutions in the same areas where the original experiences where meant to go. The shape of the surface of another person's face is encoded in the visual cortex, even if the subject might be blind. The outlines of faces can be made out by touch; letters written in Braille can be read with the fingers. The normal function does not depend on the conventional use of the sensory channels but on the degree to which a child's feeling of successfully coping with its own unfolding is being mirrored and confirmed.

"Imitation and the understanding of what goes on in the heads of the others are inextricably tied to one another," Andrew Meltzoff suggests with his "Like Me" hypothesis.[6] In order to empathize with others an infant must be like them — a body-as-feeling. Only if such empathetic behavior is granted to him in a sufficient manner can he become what he fundamentally already is. The fact that a newborn imitates others, therefore, means that from the beginning he is able to distinguish them from objects that are not alive. Only organisms (humans, animals and even plants) can be "like me." But to unfold this identity, the infant depends on other human beings who are able to see that he is not an object but alive and who are able to confirm this feeling through a gentle gaze.

For cognition researcher and philosopher Evan Thompson, who teaches in Vancouver, the experiments done by Meltzoff and Moore show: "For newborns the experience to have a self is coupled to the presence of others."[7] The self-identity is formed in a ping-pong game of communication between the infant's self and that of her caregivers. Human identity, therefore, always necessitates the presence of another individual. Identity should not be understood as solely subjective but rather as intersubjective, as Thompson believes. Every subject in reality is an "intersubject." The key to a healthy development is, therefore, found in empathy, in the possibility to experience my own emotions by recognizing their echo in others whom I know are themselves also beings with emotions. We could also express it in more generic terms: to be able to become means to be allowed to be loved. This gift, the gift of being seen, of being mirrored, is a biological necessity.

It is for this reason, not because of the supposed "primary symbiosis" with the mother, that a human is dependent on relationships of mutual exchange for his whole life. Identity always continues to unfold. On the level of metabolism, but also on the level of personhood, self can only exist as self-in-growth by continuously remaking and recreating itself. It only becomes possible through continuous exchange and transformation. The newborn must experience itself as separate and be capable of flourishing as separate, but she can make this experience only in interchange with others. She needs the gift of love, which contains the understanding that the ability to be in contact enables self, and the converse experience that only a true self makes a connection possible.

The newborn needs his emotions mirrored by the face, which is the embodied self of his mother or father, in order to develop a true understanding of himself. Their gaze allows him to become the individual that he already is. According to the degree of their empathy, their understanding and their acceptance, he will become a different person. Cognitive experiments in apes have shown that the degree of parental care even influences the anatomical structure of the brain in a crucial way. Emotions that a child has not experienced by a certain age because they have not been met with an echo in the other are deleted from the brain architecture and then are lost for a whole lifetime. The concerned individual will then always be dependent on others' feeling to experience his own. He will never step out of the infantile need to be mirrored.

In a cruel but revealing manner researchers have repeatedly shown how deeply decisive the influence of other is on self. For instance, they have examined what happens when animals are prevented from participating in the healthy interplay of visual and motoric exploration of their environment. These deprivation experiments show that if kittens or infant apes do not learn to orientate in their environment by a certain age — because they have been kept in total darkness or grew up tied to a flat trolley — then it is too late to make up for the capabilities, or rather, that part of the self that has not been developed. Cats that grow up tied to a board on wheels never learn to walk in their life. The sensory world that these miserable figures have created around themselves from only their restricted experiences is a spooky caricature of reality.

For Thompson, brain research has not taken sufficient account of how the entanglement of self and other contributes to the development of self. For a long time researchers believed that we are surrounded by an external world that is objectively given "out there" and which we can only perceive by receiving neutral information sent from it. By now many scientists suspect that not even self-identity is granted; rather, it has to be constantly achieved. They have started to glimpse that this constant manner of self-realization and self-assurance that searches for the self in the eyes of the other could make up the feeling core of consciousness.

Evan Thompson is convinced that in principle we can conceive subjectivity — our experience of ourselves from the inside out — only in an interchange with another subjectivity which answers to the self and which is "like me" and yet different. The other is always part and parcel of the self. Otherwise, self dissolves in the whole of reality. This reasoning reminds of the observation that on a material level, the organism itself and therefore, the core of subjectivity and feeling, is an entanglement of self and other.

The becoming self knows itself predominantly from the inside as a world of feeling. It can take up a more objective relationship in relation to its affects, and through this develop a coherent, mature identity, only if it is allowed to see emotions before itself in an objectified manner. To see emotions like this, however, is only possible in another body, which is a real object in space. In order to allow the child to perceive his own emotions in this body, it must be a living being: no other bodies experience and show emotions. In order to learn his own identity, the infant must be surrounded by other living beings: no other pieces of matter reflect inwardness through their outer shape.

MIRROR NEURONS: REFLECTED WORLD INSCAPE

Meanwhile researchers have decoded some of the neurobiological mechanisms by which the self pervades the other. In the late 1990s an Italian research group led by Giacomo Rizzolatti and Vittorio Gallese identified nerve cells ("mirror neurons") in the frontal brain of macaque monkeys, which fired their impulses not only when the animal made a movement but also when it observed another macaque while he moved.[8] (Later it was even found that newborn macaques imitate human facial expressions exactly in the way human infants do.) The discoverers of the effect concluded that the other's body is imaged in the brain, or rather, in the new terminology being developed, becomes meaningful in relation to the core self of an observer.

Humans also have these mirror neurons. They make up, as it were, the circuit of compassion. And this circuit is an important part of our brains. Nowadays researchers do not speak of single mirror neurons, but have rather identified a whole mirror system, which is distributed all over the brain and which does not only respond to images but also to sounds. This system is shared by many vertebrate species. It is still not clear, though, if the mirror system is a specific functional sub-system of the brain or if the mirror system rather exemplifies the normal way a living being's entire brain reacts to other beings. Maybe all brains are mirror systems. And maybe they are only capable of meaningful cognition insofar as they are mirror systems cognizing through an intimate swapping of one's place with the other's.

Through the mirror system the core self translates the existential experiences of the other, which are perceived on its outside, directly into its own feelings. The core self is symbolized as outside and by this is able to extend the sphere of inwardness. Mirroring is built into the way the existential meaning of core self structures cognition. It is, therefore, possible that mirror behavior simply is the way a brain connects inside and outside. That would mean that every living being automatically, by being alive and having a body, is hardwired for empathy. Empathy, after all, should be seen as the standard practice of perceiving self and other and not as an extra feature of some brains. This perspective also entails that our brains are always activated if we see an organism undergo experiences, be it as distantly related as the slipper animalcule of Chapter 5 or even a plant. Think of the imposing gesture an old oak tree shows. I am sure our brains resonate with it as well. After all, every gesture of a living being translates itself in the dimensionless meaning space of the core self, which is pure expression. This space is not reserved for the individual. It is where all selves meet.

Through the neural mirror systems, which inevitably manifest with the symbolic and emotional functioning of the core self, the borders between two individuals necessarily become permeable. If the self can only participate in the experiences of another being by feeling them in his own body as mattering to his own core self, then self and other somehow have swapped bodies — or all are alive as organs of one huge body, which extends through a host of individuals. Following the logic of the mirror system, therefore, we can imagine all organisms connected in a huge meshwork of experiences, which are existentially real even *before* they have obtained their explicit shape through the senses or through reason and language. This is the poetic space in which we meet other beings. It is the sphere in which Paul Cézanne, the painter, could state that while painting the Montagne Sainte Victoire in Provence, he realized that "nature is inside," and it is the dimension in which an indigenous hunter intuits the presence of the animals in the forest.

Cognition researchers do not know if the mirror system is innate or if it develops by the early imitational behavior we have been describing above. But they have found other brain centers, which are located in evolutionarily deeper areas than the neocortex, where the brain deals with a number of emotional impulses. Jaak Panksepp for instance holds that the periaqueductal gray matter, located directly above the brain stem, plays a crucial role in connecting identity with other. Here impulses which evaluate a stimulus are connected to sensory information and are joined to feedback impulses from the motor nerves coming from the spinal cord. If empathetic behavior is embedded as deeply as Panksepp thinks, this means a nearly universal accessibility of core self feelings across at least the vertebrate realm. It also means that vastly more animals than only humans and the higher mammals are dependent on the mirroring effect of the other in order to behave coherently and to have a unified self-image.

The consequences of the mirror system show that in a certain sense we already are the other. Entangled by the mirroring process between self and other a multitude of organisms becomes a sort of unity. All organisms are reliant upon one another in order to fully realize their potential ability to sense and feel and create their identities. The more dense the ecosystem, the more profound its possibilities for experience and understanding. The most impenetrable primary forest, the most complex web of entangled life, reveals the deepest insight. We can share these experiences if we become part of the ecosystem. If all beings are reliant on one another in their perception, then organisms indeed form what the French existentialist Maurice Merleau-Ponty has called the "flesh of the world," a thick tissue consisting of that which is perceived and those who perceive it, a junction of gazes, which creates reality in that middle ground where the mutual gazes intersect. What Merleau-Ponty describes as a poetic metaphysics, however, today we can trace through cognition research and understand more deeply by describing the lived reality of a human newborn infant.

These latest empirical findings from cognitive science liberate philosophers from a century-old thorn in their flesh. A typically stubborn philosophical problem has been the question of exterior experience. How can we explain how a human can know what another feels or thinks, although he only perceives the other's body from the outside? In some respect, the whole opus of Immanuel Kant, the famous Prussian philosopher, is an attempt to solve this dilemma. But the whole question is only problematic if it is tackled by logical reasoning alone. The "reasons of the heart" of feeling perception, however, always intuitively grasped what the applied neurosciences now have started to conceptualize in scientific terms. In the logic of the heart the interplay of two imaginations can never be torn apart because imagination is only possible through an interplay of self and other as mutual inspiration. There is a deep poetic reason for this sharing, one which has to do with the way a healthy identity unfolds and also with the fact that feeling expression is nothing solitary but always a metamorphosis of the whole through the individual. The whole becomes transparent through the individual and vice versa, as the French poet Paul Valéry observed when he noticed: "You take my image, my appearance, I take yours. You are not I, because you see me and I do not see me. What I am lacking is that I that you see. And what you are lacking is you, whom I see."[9] Every living body is always this crossing over of perspectives. It is inward and outward at the same time, and one only through the other.

We can prove this double relationship, which Merleau-Ponty called the "chiasm" of living perception, at any moment with our own body. Just touch the back of your left hand with a fingertip of the right. Close your eyes. What you can sense is that in this very touch, which you feel on the back of your hand, the finger not only touches, actively but is also being touched, namely by the left hand. To be perceptible it must itself perceive. And as the finger tip is perceived, it is not perceived as such but only as a change in the self-experience of the skin on the back of the left hand. Perception always means perception of oneself. And this self-perception needs another subject to be possible.

The lingua franca of which affective neuroscientists speak, therefore, is an Esperanto of the body, not of the conscious mind. Presumably all who live can understand this language. And they all can communicate with one another only through this language. We can come into contact solely through contact with our vulnerable bodies. Their proper vulnerability, therefore, is an organ of perception. And because we do not possess our body in the way we can own an object — because we *are* our body whose matter at the same time continuously passes and is renewed — we need to understand ourselves through others. Meltzoff observes, nearly with surprise: "The ability of young infants to interpret the bodily acts of others in terms of their own acts and experiences gives them a tool for cracking the problem of other minds."[10] Did we not always presage this, dear parents among my readers? What we tend to forget, however, is the second half of this truth: we learn to read our own soul only if we can see its inner tension exemplified by another living body.

IN THE ANIMALS, PLATO'S IDEAS
BECOME FLESH AND BLOOD

All life forms in nature are bodies, and these bodies are all related to some aspects of our experience. We can be acquainted with all of our embodied emotions during childhood only if we can meet them in an externalized way in other beings which we love and which we care for. This is the grand role of animals in a child's imaginary life. They are of enormous importance. All beings show their feelings — not human feelings, but the more general emotive states of being alive, of sensing joy or suffering. In many ways, what it means to be a living being in a sense stripped of the aspects of social and cultural imagination can be experienced more purely in animals than in humans. Animals are so much more directly connected to their existential needs than we are. They *are* their feelings in any moment. Perhaps they constantly live in a realm of synesthetic knowledge in which primary experience and meaning have not yet been split up.

Humans may be able to unfold their emotionality to its finest degree only if they are allowed to see the most general emotional archetypes in front of them. Animals give a shape to the desire for life. This — and not the assumed personal characters of particular types, which have been imagined by the authors of classical fables — is the deep symbolic meaning of their presence. Through an animal's particular way of being, which is engraved in its gestures, certain modes of being can come into the world. These modes become real because they manifest in bodies. At the same time they remain symbolic. We cannot describe them in their totality in a scientific, objective way. The bison's strength, the fox's leery manners, the swift motion of the swallow, the gazelle's grace and the purity of the swan are not *in* those animals themselves. But these qualities become clear through them as archetypes of possible existential constellations. We would be at loss to imagine them without the archetypal expressions of the animals.

Our distinct thought categories are possible because there exists in other natural beings a form of absoluteness of what life is about, of what the existential possibilities of reality can be. Each animal is a complicated figure, a reenactment of the specific living forces at a certain point on the gradient between the totally isolated self and the complete fusion with the whole. Each being is one unique solution to life's enigma of unifying the individual with all the rest of that which is; therefore, it contains all the grandeur and all the suffering this singular solution entails. The bodies of other beings give objectivity to every yearning, every uprising, every defeat and reveal them as a secret part of the one grand tissue we are all part of. They make it ours by living wholly through it, so that we can own the self-experience and the core-self-meanings of the whole without exhausting every possible experience. They make the splendor and sorrow of reality mine. The Greek philosopher Plato suggested that there was an *eidos* for every abstract concept, an archetype in a disembodied realm of ideas, which had no earthly existence. Plato claimed that the essence of generosity or the idea of facility existed there as the abstract epitome of its essence through which the real manifestations of the ideals became possible. But here Plato surely was wrong. The realm of ideas is not beyond, in an ideal world. It is here, in the forms of the animals. They are its guardians, ready for any self-sacrifice we ever demanded of them.

While not as important for a child as her parents, animals still fulfill an irretrievable existential role. In a certain sense, related to their role in the infant's emotional development, one could even say that they are his parents — the guarantors for the development of a self that is capable of feeling alive without developing dead zones within itself.

In today's reality, however, animals have become less and less present. Instead, the world is full of machines designed by humans. Vehicles, automatons, computers and touchscreen devices all share a sort of living magic ("auto-nomic," they are in some sense self-propelled and do not need to be pushed or pulled by muscle power), and they combine this magic with being completely at our disposal. These devices pull children and most other grown-up humans to them with a sort of magnetic undertow. It is important to understand the attraction of technical devices, and it is even more important to not dismiss them (particularly computers) as "leading the wrong way," as is often done in ecological circles. But it is also indispensable to see the categorical difference between designed devices and organisms that exist on their own accord and why this autonomous existence needs to be a major factor of our world. The irresistible attraction of machines probably has to do with the fact that they also allow for a sort of mutual exchange; tools and machines make us more capable of changing the world and hence to create. This situation generates a particular kind of embodied and imaginative reciprocity. Technical artefacts mirror a potential of our body and project it into the world. They manipulate and reorder, they share the great transformative power living beings possess. But machines do not have limits corresponding to the very specific and very fragile needs of sentient beings. They seduce us to confuse creation with control.

Also machines allow for certain forms of interaction and perception that in the long history of our coevolution with nature had been only possible through animals. But they simplify the lesson because they are controllable. They are similar enough to living beings that for a long time biologists have been deceiving themselves by proposing the automaton as the most apt analogy for a living being. But such a view not only misses its object, it also fools us about ourselves in a dangerous way. Instead of the humility of origin, machines give us the drunkenness of power.

Human intelligence is tied to the presence of other creatures. They give us the means by which our cognitive abilities, our capability of being in the world, our sense organs for reality, sharpen. They are the vehicles of our identity and our consciousness, for only in their company do we grasp that qualities and character traits are objective attributes of this world and not mere projections or whims. Animals are the guarantors that destiny is real and that it remains a force that we cannot mould with technical means. Only in their vulnerable bodies can we find our human measure, a harbor for our own fragility. Without them we would perish from immeasurable hubris. No mortal is made for this kind of Phaetonic flight of fancy with which technical power has been seducing us for centuries. This warning is written into the weak and welcoming bodies of the other beings.

Animals are like us, and they are not like us at all. For this reason they help us mature and gain a distance from ourselves. In the company of an animal, be it as closely related as a chimp or as distant as a tadpole, we find something in ourselves, which is intimately known. But what is known is always entangled with total otherness. It always carries with it an abyss of newness, which is the prerequisite of all creation. For our most aristocratic qualities are preserved in the animals — in the tiger's tense calm, in the lion's majesty, in the buffalo's strength, in the uncanny shrewdness of the fox, in the farsightedness of the owl. All these are not emotional characteristics of particular species but gestures of life itself. They are aspects of our own possibilities, which we can only recognize in their most consummate way in animals. The whole existential spectrum of a flesh and blood existence, its exultation and its despair, lie contained in the animals and only in them. Animals convert the world inscape of our feelings into visible life. They are the devices of subjectivity. To experience animals means to understand the condition by which subjectivity is possible.

Animals blend our own qualities with wholly new ones in a mixture that in the end is unfathomable, never to be untangled. "The essence of life is contained within the substance of life and cannot be 'extracted' without killing life; full transcendence, therefore, is impossible," literary scholar John Bryant noted in reference to *Moby Dick*, Herman Melville's deeply moving, cosmic novel.[11] Transcendence can only be grasped if it immediately becomes immanence, if it is experienced as transcendence-in-immanence. It is this quality which makes other beings so precious to us. They allow for our own transcendence through their opaqueness, through being bottomless and at the same time like ourselves. They show that our feeling is something totally simple, open to all living beings, open to the world and that we can understand it without philosophical reasoning — and they prove that it is never to be wholly understood.

This intertwining of something that is our own and something that is totally alien is nothing new for us. To the contrary, this experience, which demands nobility of mind as an attitude towards being, carries an old name. It is called love. Love is the only conceivable relation in which egoism and altruism can achieve a balance because one is only possible through the other. You wish the beloved to live, regardless of where she is situated, regardless of whether you will ever see her again. It is our love for this other which makes us grow, which strengthens us, which lets us be ourselves. What gives us our identity in this mature idea of an ecological practice of loving is not what we alone possess but what we are able to give in an interchange. The important point is not what the self receives from the other, but what it is able to provide in order for life to be. This is the absoluteness of the gesture of giving. It is a gesture that rejuvenates the life-giving ability of the living cosmos.

In the world of organisms in which we have evolved as one species among many, the possibility of this love grants us humanity. When there are far fewer animal species, we will be less able to truly be ourselves. "Romantic pessimism," some might respond to such an insight. But my journey through Ethiopia taught me a different thing. The last predators of Abyssinia are not only as precious as temples, sculptures or paintings, which should not be destroyed. They are a part of us, an element of our soul. Life's richness and material wealth are entangled in mysterious ways. Only those ecosystems are whole through whose grasslands and forests roam the lone top predators who have come into being over millions of years. A carnivore like *Canis sinensis*, therefore, is the intersection point of unfathomable complexity embracing not only its biological role but also its meaning in relation to all other beings of the whole system. We do not grasp this complexity, and we are unable to fully live up to it, but it holds and nourishes us with its products. The beauty and the vigor of the wolf is the power and the kindness of a blossoming life-system, which only thus, in its flourishing, warrants our human subsistence. If the wolf will not survive, then the humans won't, at least not in their full humanity. Ethiopia showed me the bitter truth of such an idea. The people already starved to death on an earth that had become stone.

In the late afternoon I stood across from the wolf a last time. This farewell in the evening light burns in front of my inner eye as if it had been yesterday. It was an experience that will fulfill me for all times to come. The wolf rushed up the hill. He dived with long leaps through the billowing Helichrysum, gracefully dividing the silver ocean of its dried flowers and leaving a silvery wake. Up he went in a straight line, up until directly below the hill's rim. There the animal halted, turned around hesitantly and looked at me for a long time with his black eyes. Through the magnification of the binoculars his gaze conveyed an abysmal understanding. He stood there, as if he was the last of its kind and, therefore, he had pity for the humans. When he veered away with a leap and resumed his track, a wave of light reflections preceded him through the dry gray plants as if they were an electrical field. High on the ridge, before the wolf vanished into the adjacent valley, the oblique sun for a second set its fur ablaze.

The next day Getachu was as good as healed again. The Lariam had done miracles. We decided to stop a car and to go back to Debre Sina together. We left our luggage in the camp. We handed over a couple of worn hundred birr notes to Demessi so that he would keep an eye on everything, assisted by his rifle. Best would be to find a truck and hop on its load area. This was one of the most comfortable ways to move in these regions. We set out on the road and started to walk, waiting for a ride. The wind caressed my hair with all the lightness of being. There was no traffic. Behind the first curve a dead honey badger lay in the rising sunshine.

About noon we met the priest. He too was walking on foot. From afar he seemed to be a huge rose-colored butterfly. He was barefoot and wore torn white harem pants. A pink cloak full of ruffles covered his head and shoulders. The skeleton of a white silk umbrella, which had lost nearly all its tissue covering, dangled across his arm. In his right hand he carried an open Bible made from yellow parchment, which was greasy and dirty. When we met, he smiled. He was young, with a sparse beard. I saw that his book was written by hand. I gave him a bank note, and he blessed me with the enormous golden cross he carried around his neck. He was full of good humor, full of understanding, bottomless gentleness. There he went, the angel of this world, dragging his torn cloak of shining embroidery behind him like a pair of magnificent but useless wings. He had flutteringly fallen down amongst us humans and redeemed us with his smile. In the distance the bare mountain peaks shone like another world as if we were imbibed by transcendence.

Part Four:
Life as Art

In this section I will be talking about beauty. I will examine why the Darwinistic idea of nature — merciless competition, illusionary emotions — is not justified. If feeling is a physical force and the expression of this feeling is a physical reality whose meaning motivates organisms to act, then we might understand living beings better if we imagine what is happening in the biosphere as, in a way, resembling artistic expression. This has another interesting consequence. Art then is no longer what separates humans from nature, but rather it is life's voice fully in us. Its message is that beauty has no function. It is rather the essence of reality.

CHAPTER 8:

Melody of the Soul

Birds are not like ideas. They are ideas.

— Paul Shephard

The nightingale hid beneath her song as if she was veiled under a liquid robe. It was early evening, and a sudden rain shower had punctured the smooth brown surface of the Wannsee Lake for some whirring minutes. The burst bowed and shook the black cherry bushes in the deserted park of the Wannsee public baths then soon moved on, leaving a sudden silence. It was then that the fluting started, following the gurgling sound of runoff from the rain, flowing like a vague answer through the silence, smooth and sparkling, close, then fading into the distance. The nightingale, invisible, sang. Notes with glassy curves rose up from the ubiquitous droplets on wet leaves and branches striking the ground. Torrents and upswings, trills and shivers, cascades of crescendos and dusky polished streams. The music rose, saturating the air, surging upwards and filling the space in a silently rising tide.

The nightingale was nowhere to be seen. She could have been everywhere, so evenly her voice flowed out of the landscape. It was as if her song changed the very medium in which it rang out, as if it blended with the air and transformed its emptiness into a chiming vault. When I must have come closer, the animal grew silent. When I waited and did not stir, she began her song anew with a volume and reverberation I thought impossible for such a tiny body to produce. A nightingale weighs less than an ounce and is no longer than a large leaf. The nightingale, it occurred to me, was nothing but voice.

From days of old, her tendency to dissolve behind the swelling intensity of her melodies gave *Luscinia megarhynchos* the reputation of being the prima donna of the birds, the fairy tale incarnation of melodic purity, moving listeners to tears. Myths and legends systematically assigned her the role of life itself. We know that all birds feed or care for their young, but it seems that the nightingale simply sings as if there was only one sex in the species, leading an androgynous existence like the legendary unicorn. The nightingale's fluting is absolute voice — more than a signal, more than a communicative device, more than an identifier of the species. If melody is the resonating form in which feeling discloses itself, then this bird's voice embraces the idea of the plenitude of existence. It is an acoustic elixir of aliveness.

The Wannsee public bath was deserted on this warm but rainy day in May. I had taken the S-Bahn rapid transit train and stepped onto one of those old pre-war cars which then, in the 1990s, still ran on the Berlin city network with their high-pitched motor whine and hard wooden benches. The train had carried me out of the concrete mass of Berlin's western city center and then over the long straight stretch of double track through the Grunewald forest. I got off at the Nikolassee station, stepping out onto a deserted platform. When the train was gone, there remained the smell of brake dust and tar seeping out from the sleepers, which steamed in the sudden sunlight. The platform, covered with old and slightly warped cobblestones, stretched out in the light, and at that moment it was as if I had the whole surface of the Earth with its infinite memories, all to myself.

The footpath to the lake was equally deserted. At its borders black cherry trees laden with white blossom clusters exploded with fragrance. Lichen and lilacs poured their flavors into the moist air, creating a sensation of spring that was physically palpable. I was happy to have left the city behind with its uniform stream of faceless people and anonymous facades. As I walked the track through the Grunewald forest I was struck by the idea that behind my back lay a whole world swelling and breathing, a world which was ready to receive me at any moment and which never asked for an access permit.

In early summer the forests around Berlin smell of dryness and light, even if it has just rained. The wet and slightly musty atmosphere characteristic of forests in general is barely perceivable here, so I wandered on in breezy exhilaration. After 20 minutes, the Wannsee, a large lake in Berlin's southwest, became visible through the trees. It was a brownish-blue color and looked like an ocean expanding to the rim of the known world. Maybe a kind of longing for the endlessness of the sea had drawn me here on a day that was absolutely unsuitable for bathing. I had no intention of swimming. I am a bit aquaphobic anyway. I paid the small entrance fee, was given a yellowish cardboard ticket printed in the style of the 1920s and was once again alone to roam the premises. The flat rectangular structures along the pale sand shore were constructed from yellow brick in the boxy Art Deco Style of the 1920s. The whole complex is permeated with the atmosphere of those years. The park was free of human presence but pregnant with the past, pregnant with potential.

Into this haziness the nightingale's voice proliferated like a drip of blood in a glass with a transparent liquid. I asked myself, How do these tiny birds manage to travel up to our latitudes every spring, even though they are tossed and churned by gales and icy air currents and so many are shot in Spain, Italy and France? They just appear in the same mysterious way that the delicate bees magically do in the spring after they leave their underground hideaways: suddenly there they are, in the heart of the crocus blossoms, as if conjured from nothing.

The nightingale sang and sang, losing herself in rising cascades that ended in sweet sobs like no other bird I know can make. These crescendo-verses give her melodies their distinctive character and make them rather similar to human music, which is also organized according to peaks and climaxes. The Bremen-based neurobiologist, Henrike Hulsch, recently showed that the memory capacities in a nightingale's brain are structured in a way comparable to our own. Hulsch thinks that a nightingale, like a prima donna rehearsing, hears herself sing, decides how well she did the coloratura and then varies it accordingly. She does this with more perseverance than any human musician, though. Sometimes the birds seem to sing ceaselessly for whole nights and days, pausing only to occasionally peck for an insect. On these spring nights the bird loses a good part of its already minuscule body weight. The nightingale shows us to what degree the world is able to become voice.

LANGUAGES OF NATURE

What drives the tiny bird to expend itself so untiringly? In the eyes of archaic peoples, who were the last heirs to the deep symbolic past that has shaped our cultural history for a million years, the birds speak as messengers from the days of creation. Creation is present in the birds in terms of a potential, which is able to give life beyond all chronological time. Many early cultures believed that it was the birds that first brought song to humans, that their voice conferred expression to our own. According to the "Metamorphoses," the Roman poet Ovid's legendary account of cultural and natural myths, it is from nature's voices that first the gods and then the humans emerge. The Orphic voice of Greek mythology manifests itself initially through animal voices. These are the voice of the cosmos, whose mysteries a common mortal cannot grasp without extensive cultivation.

Throughout all of antiquity the idea prevails that a language exists that genuinely allows access to the innermost essence of the physical world. Whoever was in possession of this mystery could shed some light into the darkest abyss of his own soul. In later centuries this primordial language was thought to be formulated in mathematical terms and signs, but in ancient times (and at certain points during the Renaissance and in Romanticism), it was found in poems and song. Poetry was the most promising vehicle for transmitting the essence of creation; song, like that of the birds, was truest to the real character of the world. Poetry and song are expressions, not equations. And it is only in expression that the emotional laws of life can unfold. This was what the animals taught the humans by singing in a time before history began.

Orpheus, the poet of antique Greek legends, had knowledge of these mysteries of creative power. Adeptly, he detected the hidden voices of nature and answered them with song. He sang so beautifully that animals, trees and even rocks were helplessly moved by his poetry. When his bride, Eurydice, died, bitten by a snake, Orpheus was inconsolable. His heartbreaking music managed to infuse so much empathy in Persephone, the goddess of the underworld, that she offered to release Eurydice on one condition: Orpheus had to lead his beloved out of the underworld by the sweetness of his voice alone, without once turning to face her or touch her. The rescue failed, as we know, because Orpheus could not resist looking back to make sure his bride was following, and she disappeared back into the death realm. In the operatic version of this legend by German composer Christoph Willibald Gluck, it is Eurydice who doubts her husband's love and requests unfailing proof of it. She touches his arm, thus making him turn around, which leads to the same tragic outcome. In both variations human passion denies song its victory over death. Suspicion impedes the cosmos from shining with the light of perfect transparency and allowing an open passage between life and death and back.

Poets continued to heed the Orphic voice. For them the nightingale's song was a thing of beauty as well as a carrier of meaning. It was proof that the principle of beauty was tied to the presence of a body and could not be detached from it. The nightingale's song kept audible the metamorphosis that living beings ceaselessly desire. The edifice of the night remained the resonant cavity of the yearning at the very core of the drive to exist, like the tiny bird's melodies in this poem by the Romantic poet Samuel Taylor Coleridge:

As he were fearful that an April night
Would be too short for him to utter forth
His love-chant, and disburden his full soul
Of all its music![1]

Another English writer of the Romantic period, John Clare, saw in the nightingale what we might call the degree zero of poetry. For him, the bird's melody was the very principle of creation become sound. Enthusiastically, he integrated into his poetry the sounds of a bird vocalizing in an apple tree in front of his window in the spring of 1832, creating a hybrid of natural and human song. He built his poetry with bird song phrases in the same way birds use human artefacts such plastic fibers or cloth to build their nests:

> The more I listened and the more
> Each note seemed sweeter than before,
> And aye so different was the strain
> She'd scarce repeat the note again:
> "Chew-chew chew-chew" and higher still,
> "Cheer-cheer cheer-cheer" more loud and shrill,
> "Cheer-up cheer-up cheer-up" — and dropped
> Low "Tweet tweet jug jug jug" — and stopped
> One moment just to drink the sound
> Her music made, and then a round
> Of stranger witching notes was heard
> As if it was a stranger bird:
> "Wew-wew wew-wew chur-chur chur-chur
> Woo-it woo-it" could this be her?
> "Tee-re tee-rew tee-rew tee-rew
> Chew-rit chew-rit" — and ever new —
>
> ...
>
> Words were not left to hum the spell.
> Could they be birds that sung so well?
> I thought, and maybe more than I,
> That music's self had left the sky
> To cheer me with its magic strain ...[2]

A CHOIR OF EGOISTS?

Is such poetic exultation only romantic love having nothing to do with reality? Is it pure sentimentality, which has been legitimately tossed into the junkyard of cultural history? Or are there aspects of this attitude that still can be meaningful, even crucial, for us today? In one respect, at least, we must pay attention to the Romantic impulse: it developed as a defense of the feeling-individual against the overwhelming determination of modernity to define reality as consisting only of neutral and value-free "scientific" principles. This cleft between feeling and rational understanding has never been consolidated. It does not belong to a historical past but forms the world we live in today, and it probably does so more radically than ever. For 200 years, since the birth of modern science, every experience, every moment bearing witness to passionate inwardness has become a source of suspicion. Human experience is thought "in reality" to be nothing more than an epiphenomenon of a blind process with neither direction nor meaning.

Biology, the science of life itself, was born from this passion to shed light on all dark traces of animism. Biology was meant to be (and still is widely regarded as) the totally unsentimental, disinterested and anti-subjective account of what "seems" to be a felt phenomenon. Even though at the beginning of the science of the living, bearded adventurers set out to map the world equipped with butterfly nets and botanizing chests, and at first did not do much more than collect and archive samplings of the incredible richness of nature; their overriding ambition was to uncover life as another "nothing but."

This perspective proposes its own sobering "truth" about the sweet sounds of spring. In the official reading, they are nothing but the soulless noise that accompanies nature's way of generating efficiency. A while ago a friend visited me. His own place is in an urban area, but at that time I was living in a rather green zone on the outskirts of Berlin. It was the middle of spring. He appeared to be almost shocked by the birds' loud song around my house. He thought that they could perhaps be a bit quieter! He felt repulsed in a way. "They're all just shouting 'It's all mine, it's all mine,'" he remarked in a bitter tone. For him the torrents of spring had nothing more to offer than this. They were a grand delusion from which it was better to steer clear.

Indeed, biological Darwinism perceives animal music as adver-
tising jingles in a concerto of efficiency. According to the accepted
view, animals' utterances obey the demand to enlarge their own
elbow space with as little power and the greatest return possible.
Nightingales sing, so the dogma goes, only because they want to in-
dicate the boundaries of their territory; in order to successfully do so,
they must outperform contenders. Vocal crescendos are their weap-
ons in the war of biological survival. The strongest female responds
to the proudest tenor, or so we are taught. With their strict rule to
take only functions seriously, to ignore every feeling and to view diverse features and behaviors solely as a means to attain the higher
goal of survival, contemporary biologists have inverted the burden
of proof even for the most blatant aesthetic phenomena. The more
strangely beautiful a body feature is, the more deeply functional it
must be. The long ornamental tail of a bird of paradise, the spotted
fur of a leopard, the spiral colors on a snail's shell must be examples
of an extra functionality, without which there would be no space for
them in the concept of efficient survival.

To integrate superfluous features in the edifice of functional
thinking, the Israeli researcher Amotz Zahavi introduced in 1975
what he called "the handicap principle."[3] According to this, the more
cumbersome a male trait is, the greater an impediment it is to a safe
life. Examples are the oversized antlers of a male deer or the absurdly elongated and spectacularly colored tail feathers of some tropical
birds. But coping with this obstacle is supposed to highlight the
true strength of the individual who carries the trait. Zahavi argues
that the more he seems handicapped, the more attractive a male becomes for potential female partners. The exaggerated trait indicates
that the individual is able to win the struggle for survival in spite
of its handicap. According to this reading, nightingales sing for 22
hours a day solely to prove to their picky female conspecifics that
they are able to survive such an ordeal without crashing down from
the branch. Thus, they are the best! Coleridge would turn over in his
grave if he heard this. While the poet was able to find life in nature's

most delicate manifestations, modern evolutionary theory nearly always uses death, the dying out of the less fit competing species, as its explicative horizon. For a Darwinist, life is war.

A more familiar aesthetics, however, has never understood the animals' voices in such a bellicose way. It did not view them as codes about individual fitness directed at competitors and copulation partners but as messages of biological perception in which something became transparent because the addressee recognized something of himself in it. For this older aesthetics, which runs like a red thread through the cultural history of the West, the voices of nature sing a melody of existence, which arises primarily in feeling.

So far only myths, rituals or the Romantic imagination have made explicit this connection between the animals' voices and human music. Here I want to elaborate on this further, but from a different angle, elucidating this relationship in the light of new findings in biology. I will start by highlighting a surprising parallel between two basic questions, neither of which has been properly answered by the current Darwinist approach: What is a living being? And what is beauty? Neither is answered convincingly in these information-saturated times in which answers and theories abound. This deficiency might very well be due to the fact that these questions are intimately interconnected and one cannot be answered without addressing the other.

In the following pages I will explore how we can understand this intimate but mostly unseen connection. And what about music? Why are we stirred by it? Why does desire find expression in sound and sound draw us toward life? Why do we seem to understand music, and why do we perceive its rhythms and melodies as expressions of our feelings? Why does music understand us? As well as these, there are other pertinent questions to be answered, more specifically in the area of biological sound research. For instance, we must ask: Are birds really trying to sing melodies or are they merely emitting calls and signals with fixed sense? What is the relationship between melodic form and bodily experience? And further: Can music be the

Orphic voice that informs us about the conditions of embodied feeling? Is music itself the lingua franca, the universal language which translates our feeling of being alive?

Let me invite you to follow these ancient myths on a provisional basis. Let us assume that a grain of truth shows up in the intuition that makes even a toddler point with a broad smile to branches vividly green with spring's first foliage where the robin calls. For her ear, which is open to the imprint of newness like a field of freshly fallen snow receiving the first traces of an animal's movement, birds do sing. The small child's discovery abides by the same poetic realism that guided the Greek sailors two-and-a-half thousand years ago, when they listened to the vocalizing sea. Deep down in the hulls of their ships they thought they heard mermaids chant when it was in fact whales flooding the nocturnal ocean space with their acoustic presence. For the sailors the sound seemed to flow from the very planks of their vessels.

THE MELODY IS BEFORE THE MESSAGE

Back at Wannsee, I walked along the water's edge, passing the flowering black cherry shrubs heavy with water, the slender birches and poplars, and the more ample forms of the willows and chestnuts. Other voices mixed with the nightingale's rolling verses. A blackbird whistled its melodic repetitions from the top of a dripping lilac. This species is arguably another rightful contender for the title of most accomplished singer. The European blackbird's tunes, with their rich and velvety texture, sweeten the evenings as early as March and fill them with a vague longing for the ecstasies of life to come.

But the blackbird's motifs have quite a different appeal than the polished cascades of the nightingale with their breathtaking virtuosity. In many respects the rhapsodic utterances of the blackbird have a much more contemporary and urban air; indeed this species began to dwell in old world cities during the last century, filling them with reverberations which previously could only be heard in high halls of trees. Some researchers believe that blackbirds have adjusted their compositions here, in the fragmented environment of the metropolis, to the surrounding technical soundscape in a kind of counterbalance to the cacophony of human chatter and ubiquitous civilizational noise, but maybe also as a sort of improvisation on it.

How lucky for today's city dwellers that it was the blackbird who moved in with them, into their backyards, onto their steep tin roofs prickling with TV antennas, singing its sweet tunes in the midst of human presence. Now, at least in Berlin, the nightingale is perched on the threshold of civilization as well. Her voice sounds on the shores of lakes like the Wannsee and in remote thickets of the vast Tiergarten park in the city center; it trills on street refuges and in verdant back gardens. Emerging from railroad embankments and the linden foliage of silent Kreuzberg streets, the birdsong imparts to the warm early summer nights the magic of some southern settlements where cultural refinement harmonizes with nature.

Down in the bushes covering the waterline another bird sang. First a jarring creak, his melody shifted to imitate the wistful verses of the song thrush, then changed again into a sequence of sounds reminiscent of the final solo of the Eurasian blackcap, finally petering out in an ensemble of babbling, chirping and gurgling noises like those of the common whitethroat. The marsh warbler, a bird more minute than the tiny nightingale, is able to imitate the voices of more than 80 other bird species. Because this variety migrates to Africa in the winter, its musical arrangements mix recognizable vocalizations of European species with the more exotic voices of their tropical cousins.

Why does a bird imitate others? Songbirds have to learn all their melodies, other than their short calls, which are innate. Young birds learn by repeating what their parents demonstrate and therefore become poor performers if they are locked away from their conspecifics early. Because musical talent in the world of birds is based on imitation, juveniles can pick up a lot of new sounds in the noisy world of everyday human culture. Some blackbirds nowadays warble like door bells and starlings, who are very gift-ed mimics, now and then compose variations on a theme from a construction site — an ornithological rap, made up of vibrating machines and concrete cutters.

Not only birds adapt to the noise of civilization. Marine sci-entists have observed that the melodies that orcas and humpback whales send through the sea have been changing rapidly over the last four decades. Instead of the psychedelic cetacean blues rhythms of the 1960s, nowadays the broken syncopes of grunge and rap pass through the waves as if the animals had adapted to the augment-ed underwater din in a way somewhat similar to how street kids in American inner cities have reacted to the soundscapes of their neigh-borhoods. The whales have kept their charming way of addressing each other with proper names, each moniker unique to that particu-lar other whale, before enunciating their actual message.

The most interesting detail concerning imitations, or rather, the sampling of new sounds not present in the biological spectrum, is neither the astonishing refinement nor the sheer range of variation. It is rather the capability of abstraction that manifests in such behav-ior. Imitating a sound is a way of mirroring or representing it and thus arranging it in a unique and novel way; the behavior indicates that the actor is experiencing a certain distance between his self and the world. He perceives the world as something that can be acted upon, according to his personal standpoint. Ethologists believe that a bird's learning alien voices proves that he is not chained to blind deterministic sequences of cause and effect. His songs show that he possesses a certain degree of freedom of choice.

Let us, therefore, together with these scientists, assume that the animal voice is an organ of expression. Their calls are utterances of a self, as minimally conscious as it may be, rather than tools for the task of out-competing others. Their voices are not proof of a consciousness tied to language but of a sensitive inwardness, which we also share as the core feeling of being alive. Melancholy and exultation — these ex-periences might be the minor and major keys in nature's music.

Usually, however, we only acknowledge as true musical performances those that come closest to the melodic productions of our own species. We experience animals as more or less remote from us by how close their sounds or melodies seem to ours. We hesitatingly concede expressive behavior to the whimpering, chattering and pant-hooting ape, all the more so since he can learn to paint pictures in experiments. Also, the fact that more than half of the 8,700 extant bird species sing is more or less acceptable, as is the fact of the whales who grunt, chirp and whine in the oceans.

But what about frogs — the way they slice away with their endless staccato already makes them sound closer to mechanical buzzing instruments than to living beings. Cows moo, sheep bleat, chicken cackle, crickets chirp, locusts shrill. Below that, so it seems, reigns silence. But we only have to listen. To the faint groan of a captive beetle in our hand. To the delicate knocking sounds a stone fly produces with its abdomen. To the 14 different variants of susurration, rustling, grunting and cracking that the American bessbug can create. This huge beetle uses sound in order to communicate not only with other adult individuals, but also with the thick white larvae, which eloquently answer the various noises their adult caregivers emit. Imagining this, I cannot but be reminded of the baby imitating the grimaces of infant researchers.

Nearly all animals sound. Insects not only hum with their wings, they bring forth all kinds of noises to demarcate their territories, retrieve their young, request parental care or warn that they don't taste good. They are not silent. They speak. They do not only talk, they sing. They raise their voices to no purpose, entering the sphere of pure music, as the males of some cricket species do. After they have achieved copulation, the ground zero of Darwinian purpose, they strike up a "song of triumph," as behavioral researchers have baptized this useless but elaborate tune.

Perhaps the reason for our ignorance is the fact that many sounds chime too high or low for human ears, or that they can be heard only where we rarely eavesdrop — in the sea, for instance. Divers know the tingle and crackle that fills their ears underwater, but few know where it comes from. Air bubbles maybe? The waves? No, indeed. The sound is made by minuscule crustaceans that hiss and grate and by fish that murmur and whisper. They rub their jaws, they grind together their fin rays, they grunt with their swim bladders. Whitefish swoosh together as a chorus when they feed. Catfish hiss, carps squeak, schools of minnows pipe, the roach belches, the perch drums.

All fish are capable of emitting sounds and hearing noises. They are not at all excluded from the world of sound but rather immersed in a universe of voices and tunes, much more resonant than the air. Acoustic waves speed for hundreds of miles through water. When humpback whales rehearse and vary their tonal sequences, the sound is in the ocean as well as inside them. They sing to themselves. And contrary to the prejudice of mainstream biology, which is so fixated with the struggle for survival, for many hours a day whales are not occupied with feeding. They have free time and they use it to sing. They repeat their unearthly songs in endless modulation. Sound can be a sign that purposeful existence is being suspended. Elephants roaming in the bush groan and rumble out of sheer well-being, doing nothing, and certainly nothing useful. There are many moments when the animals' sounds have no function; they just are.

Mainstream biology's paradigms conceive the highest tenor as the most elite fighter, but many bird species are not melodic combatants but lone dreamers enveloping themselves in a veil of song. Individuals of species like the garden warbler chatter in a special species-specific low-voice melody only when they are alone and undisturbed. They are talking to themselves, delicate and melodious and totally free of utility.

Experiments have shown that the blackbird sings most beautifully when she is alone and not engaged in a vocal beauty competition with her neighbors. Her most mature sonatas sound close to summer when the mating season is long over. When the blackbird sings to demarcate his territory, however, he stitches his stanzas together in a hasty and superficial way as if he has lost all pleasure in singing. Thus the European city and garden bird's song is at its worst at the very time when his song is most necessary according to biological dogma. Most biologists still think the only purposes for vocalization are to conquer mates and defend territory — all that is presumably needed to raise their number of offspring and thus their share of genes in the total population. But many who have taken the time to observe single bird species in depth have never witnessed the duel by duet that evolutionary biologists so eagerly invoke. Whoever listens to the birds at dusk will experience highly diverse choirs, choruses at the forest's edge, in which the voices of all songbirds flow together, blended into the sonorous liquid volume of one vast self-organizing canon.

The birds themselves may not plan such an orchestral coordination, but shortly before the blue hour, as the evening tilted towards dusk, a tonal allegro of blended voices resonated from the trees and bushes of Wannsee Lake. It was a dense ensemble, resounding pleasurably in a resonant space inside me. Magically, the fragrant wafts of linden, lilac and black cherry joined into this blend like exotic essences swirling in a dark liquid. All seemed to mean the same thing in an unfathomable echo of scale invariant expressions of aliveness.

THE CRY FOR LIFE

There is one sound that reveals the abyss of the animal soul without any remainder of purposive efficiency. This is the death cry. Even nearly voiceless beasts shriek loudly when their end comes suddenly. An otherwise mute hare in whose back the owl sinks its sharp claws gathers up all his strength to emit this last shrilling cry. In the face of this, no doubt is possible: it is all wrong to quit the realm of the living. No one who hears such a call mistakes it. We recognize the sound because in the depth of our bodies we know it could be our own.

This most macabre melody demonstrates that our inner self, our human inwardness, does not speak differently than that of any creature. We hear in it, as in an acoustical mirror, the echo of our soul, which serves no biological goal but is rather an expression of being. This last desperate cry cannot serve survival anymore because it is uttered from beyond, even from beyond bodily pain because it marks the end of it. This most painful noise is *surplus*, pure expressiveness, and as such guarantor of truth. Voice, therefore, always arises as the lightest trace of feeling as the pitch of vulnerable desire and being that longs to *be*. It does not serve utility but articulates meaning. Voice is at least as much a medium of craving for life as it is a carrier of biological functions.

Why not put it like this — any tone is a sign of a feeling and, as faint as it may be, the expression of a self that wants to sustain itself, which not only strives to survive but to enjoy the fullness of every experience. The voice may be also, or mainly, a communication channel, but it is nevertheless the barometer of an emotion, however much he who speaks tries to veil his feelings. Voice is not only semantics but also always sensual impact: flattering warmth, silvery sharpness, dim comfort or blunt betrayal. Voice proves that there exists something that it is. Voice is a physical fact in the world, not a bodiless symbol but a fleshly meeting with reality.

Voice means to be touched, to be grasped, to be clenched, every time it meets the body in the physical encounter of skin and sound waves. It is a collision that intrudes into the body as profoundly as it breaks forth from it again, as echo of its vibration. The touch of sound seizes us more strongly than does the image, which is always seen with some distance in front of us. Neurologists know that people who can hear but not see feel as if they are thrown into the midst of a roaring reality; on the other hand, deaf people, even when they can see perfectly, feel profoundly separate from the world, as if they are condemned to a life behind a wall of glass.

Nowhere does the inner experience of organisms reveal itself more clearly than in their voices. In each utterance there is a hint of the ecstasy of experience, the wonder of being rather than not being. What strikes me in the voice of another living subject is the fact that it is me who is affected by the world. Sound is a medium in a three-dimensional space, which, different from light, seizes my whole body, also a thing in space. Organs react to sound by adjusting their pulsations to its pulsating tide. The heart beats slower or faster, tuned in to different rhythms. Our connectedness to the acoustic environment is not as deep as it is for whales, whose blood flow becomes part of their songs and melodic oscillations in a liquid universe. But we are also always filled with sounds perceived and produced.

"THE OPPOSITE OF TIME"

If we translate the principles of animal sounds into what they would engender in the human world — a voice expressive of quintessential feeling, embracing the listener and revealing his soul — then we arrive at the principle of music. But our culture denies organisms the status of subjects and accordingly overlooks the expression of this status through their sounds. But this same culture everywhere indulges in artistic statements beyond any functionality. So again: Why make music?

No contemporary theory has developed a position on this that goes beyond the assertion that artistic symbols are only related to themselves and that any meaning attached to them is due to historical references to times when the accordant feelings were a cultural fashion. Artists continuously try to embody feelings. Art itself is, as Henry Miller put it, "about nothing but the meaning of life," but still, this meaning (or multitude of meanings), which has to do with body and with feeling, do not play any serious role in our view of art as a driving cultural force. We humans do not refuse for ourselves the right to beauty for beauty's sake. But we normally do not ask why we have this penchant for beauty or how it is related to the way we exist and to the world. We refuse any coherent explanation for our all too-obvious bias towards aesthetics, which is intimately connected with our experience of self and world. Feeling, therefore, has not only vanished from natural sciences, it also leads a shadow life in the images we have of our needs and ourselves. Feeling has more or less officially been exorcised from its most important resource, the arts.

The history of this denial is quite recent. Over the last half-century in philosophy, variants of so-called structuralist and post-structuralist ideas have dominated our assumptions about the meaning and the purpose of artistic experience. These attitudes are extremely hostile to the idea that anything outside the human mind possesses knowledge, although philosophers from these traditions nowadays do not talk as much about mind as they do about language or "discourse." Accordingly, most social scientists today claim that we do not find meaning but construct it, without any reference to an external world, by exploring the combinatorial possibilities of signs and symbols, words and forms, gestures and designs. Artistic meaning, this position insists, is linked to cultural trends or metaphoric fashions but has no relation whatsoever to an incarnate reality of living forms. Interestingly, this parallels the Darwinian approach in biology where the outer shape of an organism, its gestures, its bodily expressions and particularly its inner experience have no value in their own right but only in relation to an economy of survival. In the mainstream philosophy of the arts this functionalism is repeated. The fragments of sense that a culture generates gain meaning only according to an economy of linguistic signs, which do not have any significant relationship to the bodily reality of the being who expresses them. Mind or matter — viewed from a functionalistic stance, the world remains machine.

Our culture has clearly turned away from life. However we describe ourselves, we miss the center. We have lost the meaning that arises in the living body alone. We no longer know how to explain how this meaning captivates us, whether it meets us in the form of a serenade, a sonata, a jazz improvisation or a thrush's call. This double blindness towards body and beauty is not accidental. Art and corporeality stand in an intimate connection and can only be understood together. They are related to each other like the picture a person has of herself refers to the picture another construes of this same person. Only the overlay of both brings together the whole. In both artwork and corporeal existence something that is different from matter can achieve form. However, because this something cannot be detached from matter, which is its other side, it must be grasped by the senses. Only through these can it become real.

Indeed there have been philosophers who compared living beings to works of art rather than to machines. Or better, machines were supposed to be an extrapolation of organisms, and not the other way round. In the early period of European culture, when humans sensed in the animals' voices the origin of their own capacity to speak, there was a close connection between fine arts and technical arts. The inventions of Leonardo da Vinci still speak of this association. They were regarded neither as personal creations of a specific author nor as modifications to an imperfect layout of the world's hard wiring, but rather as outpourings of the cosmic creative harmony that had no author but was self-ordering or self-moving, as Aristotle had stated before. But in the wake of the Renaissance, when Christian belief merged with the idea of technical mastery, the idea became central that the world had a builder who rationally planned his works as if on a gigantic production line. The world became a technological feat. This way of thinking is still present today in the claims of Intelligent Design. The proponents of this view assert that many features of an organism (the eye, for instance) are so complex that they could not possibly have arisen through biological processes alone. Therefore, they posit a divine engineer watching over his creations. This is indeed the worst flaw of ID: a creation conceived in this way is thoroughly passive. The creator might be alive, but his work, including us, is but a dead mechanism.

What connects works of art and living beings from the viewpoint of poetic ecology, however, is exactly the fact that they are not passive creations. Both are always more than they seem in any given moment.

They cannot be entirely fathomed, and they cannot be fixed in a concept, because they are not solely rational. Nonetheless, both the work of art and the organism create their own objectivity. An artistic symbol expresses the feeling of being alive that we cannot explain in all its depth and all its detail. Doing this, the work of art itself in the end is not totally explainable. To help us understand, it must remain a riddle. And it is in the utterance and gesture of music in particular that these dimensions meet. Voice is totally transparent and inscrutably enigmatic.

Composers, critics and cognition researchers have a longstanding quarrel concerning the exact mechanisms by which melodies influence our emotions. Do the tonal sequences *represent* something, as a drawing represents a figure? And if so, what? Or are cadences and chords mere play that acquires meaning through no particular connection to the surrounding world, in the same way as contemporary philosophers think of language?

Let us have another short look at the view that was dominant in the Romantic period. This cultural epoch has left deep traces in our image of musical expression. Romanticism bequeathed to us the intuition that melodies excite us and make us feel a certain way because tonal sequences are symbols for feelings, just as the colored daubs on a canvas create the symbol of a landscape. According to the Romantic idea, a composition is the expression of emotions that had pervaded the musician when he created his piece. Historical critics of this sentimental interpretation, for instance the influential Viennese music scholar Eduard Hanslick, invoked against this view the fact that a scream of anger also expresses emotion but cannot be classified as art. And indeed Hanslick seems to have an important point. The presumed melancholy in a composition does not cause the listener to become sad but rather alleviates his sadness or indeed fills him with joy. If you accept this objection, then a tune must be more than the symbol of an affect. It seems to be more precisely the model of how one emotion is embedded in another and how all are interwoven into a living context. Music is not a symbol for a sentimental feeling but the presence of feeling experiencing life itself.

In the dynamics of music we grasp the character of a particular emotion with a precision similar to how we perceive the skeleton's bone structure in an X-ray, except that this precision refers to emotional perception, not abstract rationality. A piece of music, therefore, does not express feelings but ideas of feelings, not happiness but the concept of happiness, although this concept is offered not in a rational sense but as understanding through participation. It is a "source of insight, not a plea for sympathy," as the American philosopher and art critic Susanne Langer expresses it.[4] From this perspective, music is closer to the lingua franca of existential affects. Musical rhythms and cadences possibly echo — and, in turn, lead — the swelling and waning of organic processes. Melodic tensions such as expositions, developments, recapitulations, crescendos and codas, are analogous to the rising and waning of inwardness. This all happens at a depth and time before conscious human emotion has set in. Music objectifies embodied existence. It is a testing ground of the desire for intensified being.

But if music is the objective expression of inner life processes, then melodies and feelings exist on one and the same level. Both must be symbols of inwardness, the language of the organic processes themselves, as experienced by the living subject. Birdsong, then, is not music in our cultural sense but equally expressive of the logic underlying both. It contains, as does human music, the tonal translation of the principles of inwardness. In this, and not in the composition of its harmonies, it is related to human music. For this reason it affects us, and we can understand it. We identify the nightingale's song as a form in which subjectivity appears. It is alien to us, but it underlies the same conditions as our feeling does. The bird labors against his death by imagining life. In this sense, Coleridge's lyrical impression that "an April night / Would be too short for him to … disburden his full soul / Of all its music" could not have been more precise. The night is always too short, because music is life itself.

The French musician Olivier Messiaen pursued the connection between humans and feathered singers his entire life. He confronted bird's music like no other composer or philosopher in the last century. A great number of his compositions refer to nature's voices. At the beginning of World War II, in northeastern France, Messiaen was stuck in a trench for weeks, waiting for a German attack. Day after day during this unnerving wait, a skylark poured his spring chant into the air above him. Messiaen composed a legendary piece of music during his entrapment in the trench between life and death. In his famous *Quartet for the End of Time* there is a movement entitled "The Abyss of Birds." Shortly after the German offensive at the Western Front, Messiaen became a prisoner of war and managed to perform his chamber piece in the German POW camp, Stalag VIII-A where he was held.

For the composer, an all-approving, bottomless positivity displayed itself in the birds' voices. "Birds are the opposite of time," he emphatically noted later. Messiaen probably wanted to express that they are life that wants to be felt right here and now, and that can captivate the present moment with its plenitude, extending it to an eternity. In his later works, Messiaen was obsessed by transcribing the birds' melodies into human music and in this way to inscribe the voice of feeling. He composed by learning from the birds — not in a superficial, imitative way but by searching for the expression of the unspeakable principles of embodied existence. In none of his pieces do the birds' songs resemble that of the real species. The oriole from his "Catalogue of Birds" is but one example. This is not verisimilar music. It is not a view from the outside but, on the contrary, an attempt to grasp the meaning of living expression from within. Messiaen's works are an exploration of poetic space. He tries to transform tonal expressions from one embodied awareness to another, following the principle of scale invariance. For instance, he slows down the original tempo of the birds' tunes considerably in order to make their vocalizations understandable to our acoustic perception. There is a miraculous quietness and serenity about these pieces. You can listen to them tirelessly, just as you can to the nightingale's air.

THE ORPHIC VOICE

The nightingale's song, which is, after all, alien to any human tonal arrangement, does not only allow us insight into what feeling — subjectivity — might mean. There is something more to it, which surrounds it like a halo encircles the moon on a foggy night. The nightingale's voice reveals the whole spectrum of what above we have called the *conditio vitae*. The undreamt-of sound opens up the possibility of feeling certain dimensions for the first time and in their entirety. These are dimensions that we might not yet have experienced and that otherwise we might never experience at all. The American philosopher Ivy Campbell-Fisher remembers that she experienced certain emotions for the first time only through the presence of music: "My grasp of the essence of sadness ... comes not from the moments in which I have been sad, but from moments when I have seen sadness before me released from the entanglements with contingency. We have seen this in great beauty, in the works of our greatest artists."[5] Applied to the animal's voices, this means that because in the nightingale's soaring ecstasy, the animal's inwardness manifests as a living body, we can enlarge our own understanding of what it means to live as a body-soul. We expand our own inwardness, sheltered under the invisible dome, which the verses of the tiny bird spread upwards to the sky.

What we experience as feeling in this way, however, is not human joy or sorrow. The nightingale's song is not the emotional expression of a musical individual but of the conditions of existence. It does not refer to personal challenges but to the drama of being alive. Its themes are the tides of the organism, the rising and ebbing of the life force, the flourishing in springtime and the end, which always comes too quickly.

The poet Heinrich von Kleist has described two extremes of an objective attitude in his famous essay "On the Marionette Theatre": the blissful unconsciousness of nature and the objectifying perfection of consummate art. Kleist claims that only the most accomplished art of marionette play (referring metaphorically to artistic creation as such) brings forth a grace similar to the unconscious movement of the body. These two extremes also meet when we compare a musical composition with an animal's melody. Its body's expression shows objectivity undiluted by a personal bias. It shows objectivity, not in a technical sense but for the reason that a body is a real thing in a real world, struggling against death, which forms the ground zero of all existence and hence connects all beings. An accomplished work of art, on the other hand, is beautiful only to the degree to which it has been purified of the happenstances of personal life. Both have one aspect in common: they are necessary.

In the animal's song this necessity breaks through as voice. Its modulations are the absolute expression of the animal's existence. This is where animals are so far ahead of us. Their unconscious wisdom teaches us that something can be known even if it cannot be named. Because of the animals' lead in terms of necessity, any human hubris, any self-conceit, any feeling that "our" culture is superior to "their" nature is a tragic misapprehension.

The language of music moves our souls because our inner tides are also pervaded by the same organic necessity. The feeling of living subjects, which I have been describing in the last chapters, finds a counterpart in the harmonious relations of music. The animals' melodies are a central symbol for those unspeakable paths of feeling. They are the sound of life itself. It is this connection that renders the voice of the other so precious. Through it, inwardness can be known. It is enclosed in the animals and becomes accessible only through their living gestures: the seagulls' cries, ruptures in a storm-ruffled sky; the owl's call in the forest's night; the hissing shrill of cicadas in an olive grove at noon. These are sounds, which convey the self as a variant of plenitude, not only as the part called self but also as the whole.

In the voices of these beings the world sings itself. The Orphic song, that sequence of stanzas whose textures and tonalities, according to Greek mythology, built a stairway down into the deepest mysteries of the cosmos, does not depend on the imagination of a poet. The counting, measuring science that understands the world as a gigantic mechanism has been turning up empty-handed for far too long. Finally, now, it may discover that Orpheus's voice has always been here. It reverberates in the voices of the animals like an endless variation on the theme of being alive. The sound waves reiterating this theme travel through our body and are echoed by its liquid existence. We are a part of it.

CHAPTER 9:

The Principle of Beauty

Freedom is a gift of the sea.
— Pierre Joseph Proudhon

The rolling waves arriving and ebbing on the shore contain all that is necessary to know in our encounter with the sea. They breathe in and breathe out; they begin, and they end. The surf's fringe is the utmost edge of what we may desire. Its fleeting line translates the message, which arrives from out there, from the impenetrable body of the sea, and so offers it a shape accessible to our experience.

It has taken me a long time to finally arrive here, at the Britanny coast near Dinard in western France, where my footmarks disperse between the rose-colored calcareous shells, among violet and blue fragments and the blurred streak formed by drying kelp. The waves kindly bow in welcoming movements and thereby gently flatten the sand; they polish it incessantly through the fading momentum of blue and green ridges, which rise up one after the other, unrestrained, acquiescent, dividing the oblique white daylight into thousands of blue and green crystals. And out there, with all its facets, lies the sea.

Understanding the sea means understanding the world. The ocean's gestures and movements are a key to reality. If you have dwelled for a while close to the vast water's shores, you start to know them on a daily, serene ground. The sea exerts its life-giving force from afar. Here at this shore, the water travels many miles between the highest mark at full tide and the lowest ebb. The tidal difference is so huge that entire sandbars emerge from the sea when it rolls back. Its retreat leaves flat hills grown over with frizzy algae and blue-black mussels, which lay hidden under the surface before.

Strange landscapes emerge from the water, sceneries which extend in front of the eyes of the observer with the peacefulness of a world long past. He walks through them in amazement, without really grasping what is going on, a visitor for a limited time. The mud flats of the British Channel are an area of amalgamation, containing both land and water in their respective totality, but as something entirely unique. The tidal flats appear as a window into that other world, from which life and its laws have surfaced. Walking down the coast in the tidal zone allows directing the gaze over and over again into this hidden treasure grove of living beings. Concealed in sandy gaps under the surface, coiled together beneath the kelp's leaves, washed ashore by the waves' rhythm or hunkering in the cavities of mussel beds, unseen life forms await the return of the water. Their bizarre and beautiful shapes seem to keep the ancient promise of life's unlimited possibilities.

If you dig into the sand flat with your fingers or if you scoop a bucket of water from one of the shallow tidal ponds, you will inevitably come upon strangely formed beings of unusual anatomy: transparent green or red articulate worms with dozens of legs and shimmering scales, diaphanous shrimps that dart through the tidal pools, shining fans of spinning feathers, clams with a pearly rim of eyes lining the opening of their protruded proboscis. At the ocean, life surprises with immeasurable gifts. It is a treasure, distributed with open hands. We walk through boundless extravagance. Therefore, the oddly shaped bodies of all these organisms confront us with the crucial question: can economy and efficiency alone really be responsible for all this diversity of life forms? Does life only obey the coercion to be functional, as biologists have been claiming for decades? Does life's creative freedom only *seem* immeasurable, but in truth follows a well-calculated arithmetic? Or is there a tendency to freedom and self-realization, which becomes manifest in the biosphere? The sea gives its answer, but does so without words.

At the flat tidal coast with its rocks and islets, its tiny bays and dark reefs that line the northwest of Brittany, we can solve these questions with our eyes. The answers can be touched with our fingers, sucked in with our mouth and nose. A shoreline like the Emerald Coast, as the stretch between Cape Fréhel and Cancale is called, is composed of innumerable hybrid spaces. At low tide, its muddy beaches uncover a mosaic of transitions, of unexpected insights, of silently crackling and gurgling conversions. In front of the main beach of the small town of Dinard, at low tide the branched tubes of strange marine worms stand out, like wondrous trees populating the forests of unknown latitudes. In the tiny harbors, at low water the boats rest on their side until the next rising tide will lift them up again. At the turning point of the tides, the approaching water creeps forward on to the sandy flats, plays about the objects lying in its way with barely noticeable movement, spreads out along them and coats them with the delicate film of its presence. It washes over tiny feathers, pastel shells and decaying seaweed and suspends everything in its flow, only to release it some hours later at another place. Algal mats and wave ripples then stay behind at the shore like the ocean's own recollections, only to be forgotten again when the tide moves up anew.

The beaches in the small coves near Saint Jacut are washed clean by the waves, their blank surfaces glittering in the sun. They receive the waves with the virginity of the newborn Earth, when no plant, no creature had been created yet. The Earth seems to indulge in the arrival and the farewell of the waters as if it knew that from this gesture life followed. Patiently, unstirred, the ocean breathes in and breathes out. Here, billions of years ago, at a certain threshold of mineral concentration, the play of chance happened to come up with something unusual, which since then has been asserting itself with amazing success and a good lot of suffering. Life waited in the waters with the inevitability of a chemical chain reaction and then exploded into the blue spaces of our planet.

Like all organisms, we still carry our origin with us in the salty plasma of our cells. The ocean has been folding itself into us. The milieu of the body's fluids is our inner sea. We could hence say that our body — but also those of all the innumerable animals — is the way the ocean, a vast inside, imagines how it feels to have an outside. And when we are swimming or floating in the salty, viscous sea, we let our body imagine its inside to which we have no direct access.

At low tide in the bay of Saint Malo, which extends for dozens of square miles, the content of the ocean is stretched out in front of the dry land. The water has retreated from the bight, which reaches miles deep in to the land, generating the highest tidal range in Europe. Only the polished sand floor has remained, covered with billions of tiny shells. They have concentrated among the ripples, pastel-colored remainders of an infinity of existences. Here and there wet rocks speckle the vastness with their black hue, standing up in the air like mirages on the ocean floor. Silence reigns over this land deserted by the ocean. The quietness is multiplied by a ubiquitous tingling noise, caused by the incessant movements of minuscule amphipod crustaceans, which live in tiny burrows in the sand with entrances not much larger than a needle's eye. The small beasts come out of their tubes at low water and feed on the thin layer of organic substance the retreating sea has left on the sand.

The silence is not broken, but nonetheless starts to speak. It is visible in the form of mysterious traces on the muddy ground of the water that has been set free for some moments. If we look carefully, we can see that everything has its unique form: the innumerable snails on the moist sand; the slowly pulsating sea anemones in the small pools; the crab ducking into its hideaway beneath a tuft of shining seaweed; the marine polychaete that feels the sand in front of its hole with its fleshy palps carrying a set of vigorous jaws. Everything is teeming with hidden being. The salty flavor of the muddy sea floor proves the ubiquity of life, even if the observer's eyes cannot see a single animal. Every square foot of sand is a reactor of life, pervaded with pulsating flesh. Researchers have discovered that these sand flats are among the most productive habitats on earth. You can sense this creativity with all your senses. It is enough to walk into these lands to feel immersed in life.

In the sweet iodine smell of the mud, the aroma of creation and decay mingle. The circle of becoming and deteriorating is short circuited here. But productivity is not only enhanced in terms of sheer biomass, which is created at every moment, it also governs the diversity of particular forms that make up this biomass. The ocean in every respect is a laboratory of plenitude. The mortal wastefulness it indulges in cannot damage it. On the contrary, in the ocean destruction seems necessary in a deep sense. What an enormous potlatch the sea creates in every instant. Any prejudice of "efficiency only" has to struggle hard in the face of this blending of birth and death, and birth through death and death through birth. Often the strata in which new life thrives are layered upon the remainders of the old ones. Many beaches are not made of sand but of tiny fragments of calcareous shells. Coatings of beings are heaped upon one another and become intertwined. It seems that amidst this productive chaos organisms are indeed continuously reborn. They constantly convert into other forms. They die and instantly become food and building material for others, just like in a body where the substances leaking out of dying cells are absorbed by newly developing ones, stabilizing the whole. Beginning and ending do not seem to be separated by a real border. Deterioration at any moment can flip back to fertility, as if both in reality were the same — the inevitable Janus face of the whole.

The indolent yet at the same time somewhat hasty movement by which a barnacle's arms reach into the surrounding water and scoop feed from the gelatinous sea through the opening in its carapace illustrates the boundless resources already inside a small volume of the sea. For the French historian and philosopher Jules Michelet 150 years ago, a single bead of ocean water was enough to explain the universe by its metamorphoses. Many forms and beings already people this tiny volume. Here, we can witness the escape of living form from mere functioning without needing an explanation. In the uniqueness of living shapes lies a grandeur, which makes us immediately understand that they must have arisen through the living process of mutual creation. These forms need no usefulness to explain why they exist. To the contrary — under the impression of the obstinacy of formal variation, we recognize this freedom as a defining motive of life.

We need our bodies to be overwhelmed by the fresh wind of the sea to understand this. It is no concept, no reasoning but a call from some place already known. Algernon Charles Swinburne, the British late Romantic poet, tried to find an echo to this communion when he wrote in his "Garden of Cymodoce":

Sea and bright wind, and heaven and ardent air
More dear to me than all things earth-born; O to me
Mother more dear than love's own longing. Sea ...
More than love's eyes are, fair,
Be with my spirit of song as wings to bear,
As fire to feel and breathe and brighten; be
A spirit of sense more deep of deity,
A light of love, if love may be, more strong
In me than very song.[1]

Current biology knows little about this. It explains every variation by the supposed advantage of efficiency. The appearance of a being, the dimension in which the observer really communicates with it, all that which is real in terms of the body and the senses and which is indispensable in order to relate to a surrounding world, in biological science for a long time has basically been regarded as a waste product of utility. What immediately captures our senses and our empathy has had no value for the scientific conceptualization of living beings. For decades, evolutionary scientists were convinced that Darwin's selection of the fittest alone determined the traits that could get a hold in the marketplace of biological diversity. To this day, the bulk of biological scientists obey the dogma of adaptation and selection and attempt to explain every infinitesimal feature, each coloration of a scale and each form of a limb up to the most bizarre variation of mating behavior, solely by its utility for survival. In doing so, biologists apply the structure of economic progress to the biosphere. Any difference between the species must have brought some revenue by enhancing offspring rates; otherwise, the investment in renovation would not have paid off. (I have written extensively on the parallels between biological and economic science and common misconceptions in my short 2013 book *Enlivenment,* so here I only hint at the deep implications this alliance has for our self-understanding.[2])

Why does the mud-dwelling worm *Amphitrite* stick out short and stocky tentacles into the water to catch its microscopic prey, whereas its cousin, the sand mason worm *Lanice conchilega,* catches the same food with delicate threads supported by a glistening filigree made from sand grains glued together? Edward O. Wilson, one of the most prestigious evolutionary scientists, recently claimed that the diversity of many life forms is still a major unsolved problem in biology.[3] It was Wilson, though, who in the 1970s had rallied for the most radical program to explain structures and behavior of organisms. He founded the new discipline of sociobiology in order to ground any behavior firmly in the Darwinian economy of survival. Nothing that is is simply given, it says. All is won in war and exists only in order to survive. Wilson started the radical functionalist approach that still determines mainstream biology, although (an old man now and rather a sage) he has markedly changed his own position in the meantime.

Whoever regards nature as a survival machine, however, leaves the role of a lover, in whose eyes the beloved transmits the sense of her existence through the irresistible magic of her appearance. He takes the position of a pathologist instead. He does not believe his eyes and his other senses because he is looking for the mechanics behind the visible. He makes dissection the sole means in the search for the principle which joins the things together — only to realize that the leftovers do not reveal anything other than the principle according to which they have been disassembled: a mechanism.

Some thinkers go so far as to deny any independent reality to the body. They hold that the organism's body plan, which always springs forth in sensual presence and hence is the level on which poetic exchange takes place, is nothing more than a tool to successfully reproduce. This is the hypothesis of the selfish gene brought forth by

British biologist Richard Dawkins. It was hugely popular in the last decades, but famous colleagues like Wilson and Stuart Kauffman heavily criticize it today. In Dawkins' view the body is nothing but the vehicle the genes have invented to travel to another body: an artifact in the service of a strange egoism.

The genes, so Dawkin's theory goes, manipulate the living body and the living experience with a host of ingenious tricks in order to control their own propagation and diffusion. In the victory of a particular life form, hence, abstract information triumphs. We can see here the utmost suspicion against one's own flesh and lived reality and against the truth of one's own desires and needs that manifests through feelings. It is an attitude going back to Descartes' idea that sensory experiences are not reliable at all from a scientific standpoint. They could be deceptions by an evil demon or hallucinations. In fact, Dawkins' idea, that all experience — and all sensuous reality — is an illusion in service of something rather abstract, is a sort of reproduction of Descartes' caution. Through Dawkins' contention that only the genes are real and every feeling only illusion, we are exactly back to the evil demon the French philosopher warned against. This position is typical of the overwhelming attitude of suspicion against one's own direct experience that has accompanied modernity. And it is a pertinent example of our reliance on that rational and technical style of calculation, which makes today's ubiquitous loss of other beings lives possible. Only through a philosophy which disregards the meaningful body can the direct knowledge of other beings start to be a rare experience. This is called self-fulfilling prophecy.

NOT ENFORCED BUT BESTOWED

In such a radical anti-body stance, the splendor of organisms, though we can see, smell or feel it with our own delicate skin, also must be explained in a functional manner. The organisms' exterior is only real to the degree to which it is useful in terms of minimizing survival costs. The outward appearance of some organisms with their sometimes startling beauty in reality is supposed to be only function. This confuses partial functions of particular qualities with the reasons for their existence. Here the sequence of development is turned around: supplementary functions take over the role of explaining the overall existence of features. This reduces any aesthetic experience to zero, because in truth it conveys only neoliberal efficiency thinking.

Take the shimmering carapaces of the crustaceans, which you find everywhere in the intertidal space of the Brittany coast. These are forms with a striking appearance, but this appearance also has a role in metabolism. Their colors arise through biochemical reactions, which use the outer shell as a disposal site for metabolic waste. The white touches of a lobster's shell are composed of ureic acid that the body gets rid of this way. These forms indeed have a function. But independent of this fact, they are still something visible and beautiful. Rather than reducing form to function, one could observe an intricate entanglement of purpose with sheer exuberance. We do not need to reduce what we see to either side. We witness an interpenetration of spheres in which neither can be reduced to the other.

Take another example. The fiery splendor of the long delicate tentacles that a jellyfish tugs along through the water is exclusively interpreted as warning signals to avoid being devoured by a predator. Here the assigned function might even play a less important role than the sheer fact of appearance and exuberance. What are warning colors in the dark water for? Many marine animals defy purpose through their remarkable colorings. What about the flaming red of deep ocean shrimps? By far, not all organisms' traits make it easy for the biologist to find a motive for their supposed efficiency in the game of survival. And precisely the miraculous beasts from the gloomy abysses of oceanic night, less seen than any, are those painted in the most spectacular colors. They are squanderers who do not capitalize on their useless sumptuousness but seem to burn it away in the dusk of the sea. The inhabitants of the dark ocean spaces are the most urgent reminders of an autonomous force that lacks all functionality; in the silent darkness it manifests itself by its aesthetic power and gleams even without any observer, any mating partner to be found. They all are like the invisible layer of mother-of-pearl inside a seashell, which in the sealed casing shines forth with its glow of an incessant sunset.

Some biological scientists in the last decades have tried to find examples for a different view of the meaning of visible forms. They have argued that forms are not side effects. After all, the shape of a real body is the only dimension we can experience, in opposition to mere abstract concepts. Also we are visible and sensible bodies in the first place. The aesthetic realm is what unites us to others in a direct, experiential way. The late American evolutionary biologist Stephen J. Gould has made a convincing case against the "efficiency only" explanatory model in biology. He has illustrated his argument with examples from the arts and particularly from architecture. There as well, not every detail is useful or follows a purpose from the beginning, but everything develops from everything else through play and experiment.

Not all functions of artistic elements have been invented due to a particular goal — far from it. Many of them come about only in the course of a work or of art history. They have acquired their relevance in a secondary way. This seems paradoxical at first glance but it isn't, or rather, the paradox opens up creative possibilities. Gould uses architecture as an example. Here certain features, which are inevitable for the static survival of a building, often engender other constructive elements, which have not been planned for in the first place. They tap into what the biologist Stuart Kauffman, who invented the theory of autocatalytic networks we talked about in Chapter 3, calls the adjacent possible: stuff that becomes a real option if you turn the screw a step further but which you never dreamt of before, like using connected computers (an efficient, planned design) for the vast explosion of creativity on the World Wide Web (something that just came about).

Basic necessary architectural construction elements engender a host of new features that were not planned for at all but that can become more important than the underlying original structures. The pitched roof, for instance, whose function was to divert rain more efficiently, gave rise to the possibility of having a cozy attic parlor or a cool loft. A comparable surplus is provided by the spandrels in a wall resting on columns, as Gould and his colleague Richard Lewontin argue in a paper that has become famous as a case against total reductionism.[4] The spandrels are the approximately triangular spaces extending from the heads of the columns to the lower base of the wall. The architect has "selected" for the shape of the columns, their spacing, the curve of the vault. The triangular spaces remaining, however, are an add-on; they came about through the other factors. They were not selected for, or even chosen, but became structurally necessary. But it is exactly in the spandrels that some of the very important works in the history of sacred art have been deployed, like the evangelist's mosaics in the Saint Marco Cathedral in Venice. The pillars carry them but step back behind them. A Darwinian would have a bad time if he had to explain the evolutionary sense of the spandrels alone. But he would try.

Serious Darwinians relate every detail of a living being to its utility. But what if the most expressive features are just born essentially as useless additions, as playful poetry on the blank pages, which evolutionary pressure has left unused? Maybe this is what fills the bulk of the book of life. It is possible that a host of outstanding morphological features of organisms are such spandrels in Gould's sense. The building principle that made them possible, like a scaffold, follows the principle of efficiency, but their detailed execution shows the ecstasies of freedom. Particularly, the abyssal luxuriance of sea creatures makes this idea seem nearly inevitable. Maybe the gelatinous grace of a transparent starfish's larva has no specific survival value apart from that sole basic one of spreading out a surface in order to hover over the bottomless waters.

Or let us look at another example. Take the tentacles of jellyfish and marine worms, the tiny arms of barnacles, the mucous filaments unfolded into the water by many salp species, all the endless variation of rods and nets thrown out into the sea by its creatures to feed on the incredible number of invisible microscopic bodies afloat in the water. These apparatuses have one function, namely to fish for tiny foodstuff. But they can do this by any design. And especially in the ocean it seems that indeed any design has somehow realized itself. We, therefore, must ask ourselves: Is not the glistening of all these perceptive and perceivable worlds by which the most inconspicuous animals extend into the drifting universe of the oceans the same kind of addition for free as the dew sparkling on spider webs in the early morning light?

But although this freedom is particularly well pronounced in the oceans, it is not only here where we can observe it. It is everywhere. Filip Jaroš, a Czech graduate student of biology, recently did a comparative survey of the adaptive value of coat colors of big feline predators like leopards, jaguars and tigers.[5] The spotted and striped patterns are supposed to have an adaptive value in providing camouflage to their bearers, so that these can stalk prey more efficiently. But by careful comparison of dozens of experiments, which were supposed to prove this adaptionist idea, Jaroš found not a single one corroborating the functionalist assumption. Rather, it became clear that the coat color has no function at all. In fact, big cats with their conspicuous yellow ground color with highly contrasting black spots or stripes are not at all suited to fade into the environment structures of leaves, shadow spots or vertical grass blades (in the case of the tiger) as biologists had always thought. They are, instead, highly visible. Put an orange animal into a dull green setting, add black stripes and flashy white cheeks — and you get the most highly visible contrast in the landscape ever. And this was indeed what Jaroš found. A tiger or a jaguar could have *any* color. They can afford it. They are so powerful that they can do without camouflage. They will just soar in and take their prey anyway.

This reminds me of the beautiful lines the US novelist Cormac McCarthy wrote about the relationship of a young man and a wild wolf in his *Border Trilogy*. At the end of this episode the wolf has died an unnecessary and cruel death inflicted by humans. The hero of the novel takes the animal out into the hills and ponders his dead body at the campfire. In his imagination he can see the animal "running in the mountains, running in the starlight where the grass was wet and the sun's coming as yet had not undone the rich matrix of creatures passed in the night before her. Deer and hare and dove and ground vole all richly empaneled on the air for her delight, all nations of the possible world ordained by God of which she was one among and not separate from ... what cannot be held, what already ran among the mountains at once terrible and of great beauty, like flowers that feed on flesh."[6]

In fact, as biologist Geerat Vermeij has observed, the grand predators, often hailed as highly efficient killing machines, are not efficient at all. As warm-blooded animals they use up over 90 percent of the energy from food just to keep their body temperature at a constant level.[7] These top predators are rather like the huge gas-guzzling cars of the 1970s — not efficient at all, but having pretty nice torque.

All these considerations are equally valid for typical human qualities, whose supposed survival functions orthodox Darwinians struggle to explain and, therefore, often dismiss as epiphenomena. But surely the human species has gained some qualities during its evolutionary history that are necessarily related to others, which might have been selected for and which the latter made possible. Perhaps even our most species-specific human characteristics, such qualities as our self-consciousness, our restlessness and our ability to see a transcendent meaning behind things, do not exist because they made survival more probable. Rather, these qualities are inextricably linked to the existence of a highly intelligent and extremely flexible being. We just take nature's inherent characteristics to extremes so that they become highlighted in a unique manner, clearly showing the imaginative surplus that is inherent in embodied existence. Our traits, therefore, are not to be causally separated. We cannot examine them one by one as autonomous modules in relation to their supposed single utility.

Evolutionary psychologists, however, in opposition to these observations, insist on the selection-based origin of each distinct human character trait. Every single psychological structure, they hold, must have outcompeted rival features on the battlefield of survival. The American evolutionary psychologist Steven Pinker believes that our mind consists of particular elements or behavioral modules and that these all have been selected for separately by evolution. The brain, so Pinker thinks, is a toolkit for useful features, which can be combined freely, a kind of Swiss army knife with a lot of different blades and utensils for specific purposes. Our particular behaviors and the accompanying emotions have been selected by evolution for the uses they are related to or at least have been in our prehistoric past.

Therefore, scientists try to relate actual feelings we have to functions that could have been useful for the genus *Homo* in a past natural environment. The feelings involved in love, in such a view, are caused by a particular hormonal status, which through making us feel good entices us to the huge efforts of parental care. According to evolutionary psychologists, human erotic ecstasy is meant to keep the parents constantly committed. In an analogous fashion, to sociobiologists adulterous behavior conveys an evolutionary advantage for the male by securing a higher impact of one's genes in the whole population. Some thinkers have tried to fit traditional Western morals into this functionalist frame. For them, altruism "in reality" is a hidden calculation in which the one who moderately shares with others in the long run fares best.

Sociobiologists also assume that our language has a well-defined survival value in this evolutionary toolkit. Words and concepts have been efficient devices, they suppose, to outcompete rival groups with fewer linguistic abilities because communication facilitated the coordination needed for hunting. Whoever can name the prey, they hold, is able to kill it more efficiently. From this dissecting perspective, however, not a single human emotion remains real in its own right. Everything stands for a supposed, and more real, biological purpose. What seems most real to us from an experiential stand-point, from the scientific perspective becomes most suspicious. What we perceive inwards in all its absoluteness — our own joy, our own pain, a deep bond to a particular being, sincere commitment — for the objective gaze of science is nothing but the daydream of a survival machine.

But can we not imagine everything in an entirely different manner? Could not our human language, especially, be one of those spandrels to which Gould referred? A residual of other biological developments not needed to survive but structurally necessary as part of the development of our big brains? A trait used not to maximize utility but arising as a surplus high up in our heads, like the empty attic, which some day is remodelled into a philosopher's studio? Then the situation would be rather different. Only when the capacity of language arrived (as a spandrel) did the early humans start to explore it, not necessarily to enhance their fitness but to enliven their experiences. In the end, this model suggests, we cannot separate at all that which is supposed to be "only" useful from that which is "just" experience. Felt sense is interwoven into the most basic layers of functioning.

Such a theory of human characteristics as an arbitrary surplus dovetails in a convincing way with another well-known constellation. Many of our most useful traits are deeply ambivalent. Our social sensitivity, for instance — this typical primate heritage having become super refined in *Homo* — can be used not only to benefit a group, but also to destroy its cohesion. Our capabilities are all marked by the necessary risk of failure, which rises with growing complexity. Every refinement is always a threat to its own unfolding. Therefore, the evolutionary surplus, which hitchhikes on useful qualities like a parasite thrives on others' bodies, can be used as an explanation of many of the chasms within humankind at the same time as it is the necessary structure underlying our most particular achievements. The domains of the merely possible, the unused areas that arise by growing complexity, are not meaningful as such, nor are they good in any a priori sense. They appear as secondary qualities of other traits and, therefore, they can bestow grandeur in the same manner as they can lead to cruelty. They can engender ingenuity but infamy as well. Big brains and grammar have brought forth culture but also slavery they have provided for scientific bravery; as well as for self-destructive disturbance.

This is an important insight. The prejudice against a brutal, untamed state of nature whose deadly threats need to be constantly kept in check has marked our civilization for centuries. But in opposition to this, man's biological heritage does not stand in his way because nature is cruel and gory. Nature inside of us does not impede us because it is raw and untamed and we are not. Nature is as ambivalent as we are. Our humanity is our species-specific expression of a profound antinomy. It gets in our way because nature's heritage is not a perfect arrangement of useful qualities but a host of many options. The conflict is not between human freedom and nature's determinism. Nature itself generates freedom, which carries a necessarily conflicting potential. To put it differently: it is not the ape in us that makes us brutal but rather, because we profoundly still are great apes, we are able to choose from a broad spectrum of behaviors. Every degree of freedom that is added introduces a greater potential of creation and thus of suffering.

In reality, this whole context again shows life's irresolvable pact with death, which we discussed in Chapter 2. Humans are living their particular version of the tragedy and the ecstasy of the world's poetic creativity, which is possible only at some price, and in the end the cost is so high that it demands the self's own destruction.

ANYTHING THAT WORKS IS WELCOME

Barely anything conveys this experience as intensely as the ocean's fringe with its salty-smelling elation, with its constant marriage of birth and decomposition, where mass-dying and the insatiable will to self-expression incessantly embrace one another. When we look at the delicate forests of the sand mason worms, whose transparent appendices sway in the tide, we cannot help but feel the primacy of sheer expression. When we observe the drifting snow crystals of billions of tiny copepod crustaceans drifting through the green of coastal waters, we realize that there is more than functionality and more than the survival perspective of single individuals. Are not the medusae silent images of the sea itself in their blue transparency? Are they not the essence of what it means to be water? Are not all these forms the epitome of free unfolding and dreamy self-sufficiency and thus a way through which the environment in which they have arisen expresses itself? They are a beautification that can reveal the spirit of the underlying architecture of the building far more intensely than an abstract plan ever could. Appearance is shining forth with sheer being, with real presence.

Unlike the flowers in a meadow, we cannot say that the sea anemones, the marine flower-animals of rocks in the surf and reef beds, have become colorful through an arms' race for the most attractive insect feeding site. The anemones are colorful because they are. And it cannot be true that their colors exist only in order to convey a warning against their stinging cells, as there are less conspicuous forms with dull colors that are equally poisonous. In the sea anemones, another principle becomes transparent: animals appear as a speechless interpretation of their habitat. It is an interpretation of the functional dimension through real bodies. It is a corporeal reality that is inextricably coupled to sensual perception and not an abstract analysis. This corporeal reality has no ulterior purpose. It is purpose in itself. Its particular shapes testify to the conditions under which the organisms live, but they do not play a functional role alone. Life is a spandrel in the building of reality, illustrating its stunning architecture and expressing its deepest ideas. It shows something that is more than the sum of the parts and more than any functional purpose aims at. This something is not part of any functionality. It corresponds to a gleeful yearning that was there in the beginning, and it becomes real by instilling an equal yearning in anyone who encounters it. Also in this, it is analogous to a work of art: it is material shaped by vital import, which refers to a meaning beyond the pure materiality.

But this "beyond" is not any rational purpose. It is what happens if all purpose is suspended, as in the greatest works of art, which cannot be exhausted by understanding them according to a single meaning that the artist presumably has "expressed" in the work. The most achieved works rather become expressive on their own account and display something the author only vaguely knew. The sole way to get closer to this something, which the artist sensed but did not wholly understand, is to create an echo symbolic of that feeling, which, however, is not a solution but rather a new complication. A good answer echoes the problem by transforming it into a new question. "A good question," says the wilderness mentor Jon Young, "must be made into a quest." This quest gives an answer only by becoming expressive of the vital structure of the question.[8] In being an answer-as-a-question, a good work is expressive of life. It is a hologram of the living whole, in the same way as the organism is equally a focus through which the whole becomes graspable, although at the same time it always remains hidden in the living being's body. In his book about the ocean, Jules Michelet, the great French historian of the 19th century, gave an example of this holistic vision. He described how he was transfixed by the degree to which a fish, through its body features alone — in its bluish-gray colors, in its smoothness, covered with slippery liquid resembles the element it swims in.[9]

During the Renaissance, people came to understand nature more and more as a subtle mechanical instrument. Scholars were mesmerized by the observation that organisms seemed to be more finely wrought than even the most delicate watch imaginable, which at the time was the summit of human mechanical craft. In a poetic ecology, however, the most prodigious aspect of nature is the freedom brought forth by its members, the endless potential of creativity and the innumerable individual solutions to the drama of existence. The Swiss biologist Adolf Portmann, therefore, was convinced that life's final inner purpose was not to win and outcompete others but to be simply visible at all, to display itself by its endless distribution into innumerable individuals in order to show that something *is*. For

Portmann, life was a tale-telling art for art's sake happening on a cosmic scale: a giant play to allow for the expression and presentation of self. Every living being for Portmann was a kind of individual distillate of the world, which was able to bring this all forth.

In his landmark book *Animal Forms and Patterns: A Study of the Appearance of Animals*, Portmann assembled a comprehensive compilation of animal body traits, which do not have any selective advantage but arose as non-functional concomitants to more purposeful features.[10] Often these forms are explicitly aesthetic but quite frequently they are not visible to any eye at all. The pattern inside the shell of the chambered nautilus, a marine cephalopod mollusc; the shining hue of a huge shrimp in the dark abyssal sea; the colorful splendor covered by the satin upper-wings of big moths, which in the nightly dark nobody is ever able to see, neither predator nor mating-partner — all this seems to be arabesque, ornament, fancy and only through this can be creation — expressions of the fact that there is endless freedom where life can realize with impunity whatever enters its expressive imagination.

This sheer exuberance of an animal's appearance defends itself against being understood as a mere means for the sole purpose of survival and, therefore, nothing in its own right but, rather, an illusion. The seemingly useless feature tells something about the underlying meshwork it is connected to. It is an elucidation, a commentary, a painting that displays a more concentrated content than a fleeting gaze could convey. Life enhances reality. It intensifies creation. And in this regard it is, paradoxically, the most useless features that tell the most about a potential that lies asleep in the depths of the living. Its visibility's only purpose is to reveal itself as such.

Nature, we can infer from these thoughts, seems to have some sort of sense organs for perceiving itself through its inhabitants. To leave no doubt here, this for sure happens by way of evolution namely, through a gradual unfolding of diversity. But to argue that the warrior's brutality, the continuous pitilessness of the imperatives "never enough" and "try to be better" is the only driving force and the final expression of the biosphere, seems a tad naïve in light of the drifting world of wondrous marine creatures. Should we not rather admit that it is the other way round? We cannot help but ask this when faced with the bizarre forms of marine life amid the neat efficiency calculation. Is it not far more probable that in natural history everything is permitted that reveals itself to be doable at all and that is not in contradiction to the laws governing matter? Only the most absurd assemblies, like a fish without gills but with feathers, may really never be selected. Everything that is not a blunt contradiction of reality, however, can.

Let us invert the perspective to see more clearly. Instead of claiming that every present form has outcompeted all possible contenders, should we not say that it was not too outlandish to *not* be possible? Only too often life is not represented by the smallest common denominator but rather by the error in the biological reckoning. Why whimsical arabesques like the common seahorse's indignant snout and the male steed's pouch in the front of his belly for gestating his bride's eggs? What complicated ad-hoc explanations the orthodox Darwinian has to invoke to explain the existence of such idiosyncrasies among ordinary functional "standard" fish? Is not the seahorse, particularly, an expressive whim to be explained only through the possibility that in the end nothing really spoke *against* its construction? The creative biosphere simply was allowed to play with such a form in the protective garden of the sea's lush grass meadows that formerly covered huge tracts of the coastal oceans. Why did such a difference of forms evolve in the first place and not a simple universal fish? "Why not just one single flower?" the philosopher Albert Camus asked.[11]

Today, complexity researchers have discovered a host of biological processes that are not controlled by genes and thus are not able to be selected as "fit" or to be discarded as "unsuitable." The tiny pores that a plant uses for respiration on its leaves' surfaces, for instance, are not distributed according to the greatest efficiency brought about by selective pressure. The pore pattern rather self-organizes in each growing sprout in a very simple but highly regular fashion. By the same self-creating freedom the spiny coats of crustaceans grow, the assembly of coral reef polypifers forms, the pattern of a catfish's spots develops and the coat design of big cats is produced, as we have already seen above.

Also here, "self-organization" is the key word for a process that does not know a specific direction but nonetheless always arrives at its destination. Physicists, chemists and biologists are starting to believe that the tendency of single processes to overlay, to mutually influence one another and to build up complex patterns is one of the basic principles of the universe. Weather phenomena feed back into the ordering structures that have brought them forth; chemical reactions through response and reaction produce permanent reactive cycles; and as we have seen in Chapter 3, the most complicated feedback loop of all, the chemical laboratory of the biological cell provides its crystalline order only through the incredible diversity of the millions of substances assembled in it. All these forces provide for diversity — but they never obeyed any Darwinian selection. They are directionless, playful and inebriated by newness. It is possible that Darwinian selection could chose among several variants of self-organizing systems, but it was not able to govern their appearance. The way these reactions function has to do with the structural possibilities of reality. Their shape is not won on the battlefield. It comes as a gift. It is "order for free," as the biologist Stuart Kauffman convincingly has shown in a host of empirical models.[12] The cognitive scientists Humberto Maturana and Francisco Varela, therefore, conclude: "Evolution is somewhat like a sculptor with wanderlust: he goes through the wood collecting a thread here, a hunk of tin there, a piece of wood here, and he combines them in a way that their structure and circumstance allow, with no other reason than that he is *able* to combine them."[13]

THE FOURTH LAW OF THE THERMODYNAMICS OF LIFE: DIVERSITY SHALL BE!

The American complexity researchers David Depew and Bruce Weber assume that Darwinian evolution might be a special case of a vastly more comprehensive scenario of self-organization. "Self-organizing systems can only be influenced to a certain degree by selection," the two theoreticians claim.[14] Their colleagues Daniel Brooks and Edward O. Wiley justify this idea on physical grounds.[15] They have developed a theory of why living beings — single species but also entire ecosystems — strive for growing complexity. For many generations of researchers this desire for complication and density wherever life occurs has been an enigmatic exception to the laws of thermodynamics. Organisms increase the degree of order in the world — whereas the famous Second Law of Thermodynamics posits that the opposite must be true. In the whole universe, order decreases, and entropy, the lifeless state of plain and shapeless dispersal of matter and energy, grows. More than a hundred years ago the Prussian scholar Rudolf Clausius had formulated the equation proving that this kind of dull disorder was inevitable as soon as enough time went by in the universe. We all will die the so-called heat death, as the entropic dispersal was called at that time, physicists concluded — a mathematical conjecture that nicely fitted into the *fin de siècle* forebodings of that time. The message was plain — in the long run everything will decay, no matter how well it may seem to be going right now. There is not only one law of thermodynamics; at the moment there are three on which the scientific community has agreed. And none of them is very much in favor of life's caprices. Rather to the contrary. For thermodynamics, life is a blatant violation of cosmic laws. With tongue firmly in cheek, Stuart Kauffman has translated the three laws and their wholesale anti-organic notion into a language that is easy to grasp: "1. You cannot win the game. 2. You cannot break even. 3. You cannot quit the game."[16]

For a long time it seemed quite enigmatic that organisms do seem to win, at least for the limited time of their individual lifespans. Living beings obviously are able to steer clear of dissolving into inert matter. Instead they multiply, create ever more complicated ecosystems and obviously swim against the tide of physics. This fact has been a puzzle not only for evolutionists who tried to understand how this diversification happens, but also to physical and chemical scientists who pondered the seeming contradiction life poses to physical laws. The most famous result of these musings is the small essay "What is Life?" by the chemist and quantum researcher Erwin Schrödinger, in which the author missed the fact that organisms are an activity not only informed by but also informative to matter.

The systems researchers Brooks and Wiley try another approach. They remind us that Clausius's oracle of the "heat death" only holds true for closed systems like the universe as a whole. Organisms and everything assembled by them, like an ecosystem or the biosphere, however, are open to the world through the steady metabolism they maintain with their surroundings. Through this openness, living systems make it possible for structural complexity to increase within them in the same way, and parallel to, entropy. With this shift of perspective, the two systems researchers take seriously the categorical differences between physical and organic systems. They stress that living systems are open, not closed as machines are, but they still overlook that this openness introduces the factor of selfhood into the world. This factor is nothing material. It is the overall sphere of meaning generated by living systems — the diversity of structures, forms and behaviors, as well as the diversity of experiences. Brooks' and Wiley's ideas support a "thermodynamics of diversity." The researchers themselves think that by this shift in emphasis, which follows when we stop seeing organisms (and biospheres) as closed systems, the bleak Second Law converts into the principle of natural history. Behavior that follows physical laws and growing complexity goes hand in hand rather than is a contradiction.

Stuart Kauffman even proposes a Fourth Law, which he half-earnestly states as, "The game becomes more and more complicated and new rules are emerging constantly."[17] The new law thus describes the fact that the world explores ever more structures *-and that these structures are not entailed in the basic conditions present at the beginning. They cannot be reduced to any scientific boundary conditions; they have to be lived. The game of life has to be played in order for its full depth to be explored. And by playing it, meaning inevitably increases.

Some researchers even think that the organisms are catalyzers to speed up the journey to total thermodynamic equilibrium and peaceful entropy. The complexity that living beings build up is a fertilizer for decay. Organisms could be thought of as tiny swirls or eddies that through their complication of pure physics work towards a smoother balance of energy in the universe. Seen in this light, living beings might even necessarily follow from the requirements of thermodynamics. They are the higher gear, which entropy automatically chooses to reach its inertia goal once this add-on is available. But that would mean that organisms, far from challenging physical laws as scientists thought for a long while, might instead be a factor without which physics' picture of our cosmos could not be complete.

One consequence of this point of view is that the rise of complex structures and therefore, the increase of diversity in our universe is not a miracle but on the contrary, extremely probable or even necessary. Thus, the American physicist Erich Schneider claims that like many spectacular physical phenomena that are difficult to compute — for instance, turbulence between two distinct layers of air or water currents — life is a phenomenon that unfolds in relation to a concentration gradient. And within this gradient between chaos and order it does not behave fundamentally differently from all other physical systems, which are provoked by such interfaces to form complex self-organising patterns. Their beauty then, in truth, is nothing other than an inevitable structure formed by the system during a process, which as a whole can be described without contradicting the Second Law of Thermodynamics.

The self-enhancing abundance of the biosphere, then, is the way by which heat death can be achieved most directly. It is analogous to the highly ordered eddies and convection cells that arise if you mix two colors and that help to distribute the pigments more regularly. Here, astonishingly, the unstable and transient figures with their high aesthetic value are the simplest way to reach the goal of even distribution. The finished mixture, containing no pure traces of the two original colors, is highly entropic. You would need infinite time to unmix it again, and even all of Cinderella's helpful pigeons would be of no use. Paradoxically, complicating things in the long run means keeping things simple.

From here, arabesques can be understood as a necessary detour to the central focus, a worthwhile whim. Or is the journey already the destination? The seeming detour, in addition to the fact that it accelerates and stabilizes a system, brings another aspect into the game: it is reflexive. It is a new variation of the old common theme, which interprets its deep motive one more time, a mirror maze of shapes and forms. Physicists like Weber and Depew or Brooks and Wiley may have discovered that the processing of energy and the expression of meaning are not necessarily separate from one another but, rather, entangled.

BIOLOGICAL BECOMING: A DRIFT LED BY CHANCE

What we have observed so far shows that the phenomenon of life is embedded in a context that encompasses Darwinian selection but cannot be reduced to it alone. Most certainly, competition, pressure and scarcity do play a role in directing the process of becoming. But the structural necessities of self-organization and the factor of chance are equally partners in this game. They introduce a tendency to exuberance, indulgence and craziness, which seems utterly uneconomic and so totally distant from the capitalist attitude of modern Darwinism. Only space is needed, space and time: the vastness of the sea and its boundless rhythm. Only silence, the calm of millennia, is needed to let the jewels be polished, the gemstones of becoming. Like a species of woodpeckers on a tiny New Zealand coastal island where the male chops a hole in a tree trunk with its robust beak, but only the

female can retrieve the insects at the bottom of the hole with its totally differently formed bill, long, slender and pincer-shaped, to feed her
mate after the work is accomplished. Both sexes of the species were
totally dependent on each other and incredibly susceptible to disturbances. When humans came to the island, the bird quickly died out.

Evolutionary scientists have long grasped how rarely the selection principle Darwin has erected falls into place in reality. In nature
the situation, which inspired Herbert Spencer to coin his famous
verdict of the "survival of the fittest" that Darwin quickly incorporated into the formulation of his evolutionary theory, does not exist in
this blunt form. What Spencer described was rather pretty much the
everyday reality in Victorian England. It did not define nature, but
a historical moment in human society. The atrocious excess of poor,
ill-fed, uneducated workers dying like flies from unsanitary living
conditions and the dramatic scarcity of the most important resources of food, housing and hope are not typical of nature. Nature cannot
not be grasped as being solely "red in tooth and claw." The struggle
for survival, which biologists see in nature, takes place in human society in the first place.

In biological reality death is real, but the pressure it exerts on a
population is not that dramatic. From an evolutionary standpoint, it
is chance events that nearly exclusively lead to the creation of new
species. It is not pitiless competition among species and individuals within a species that generates a new strand in the web, but the
peaceful separation of one breeding line from another that engenders evolutionary change. High competition, in contrast, allows less
newness. It causes the ubiquitous repetition of the same.

It is amazing that some biologists, trying to be faithful to the theories they have learned, do not take into account the most obvious
observations. They would have realized that one of the most famous
examples of the validity of Darwinian selection, the adaptive radiation of the Galapagos finches, stands not for the power of selection
but rather for its opposite. On the Galapagos Islands off Ecuador,
which Darwin visited on his voyage on board the research vessel

HMS *Beagle*, a single species of finches a long time ago had managed to settle on the until then birdless island. This species separated into several lineages that fulfill diverse ecological roles. Some evolved robust bills like woodpeckers'; others, slender beaks like insect eaters sport. A microcosm of ecological niches developed.

What he observed on the Galapagos functioned as one of Darwin's major proofs for his idea of descent with variation. But it does not hold so well for selection itself. Obviously, when the birds first arrived there was no competition at all on the islands but a totally free space into which the species could project its imagination of the biologically possible. They did not carry the pressure of fit or unfit on their shoulders, but rather had the world at their feet. Certainly, the Galapagos finches are in resonance with their ecological niches: animals with huge beaks eat big seeds, animals with tiny ones small insects. But this situation developed due to the rewards of discovering new possibilities of existence, not by competition for more efficiency. New add-ons that worked nicely gained space. The ecological depth increased.

Galapagos-based researchers have observed that variations of selection pressure do have some slight consequences, indeed, but until today these have never piled up to form a new separate species. When severe drought reduced the availability of certain big seeds for several years, some species slightly reduced their average beak size, as obviously more individuals with that trait managed to thrive and raise offspring. But after the rain came back, these variations quickly became insignificant.

The Galapagos Islands are a good example of a phenomenon influencing the formation of species that biologists call the founder effect. This effect is always at work if a few individuals are cut off from a larger population and need to start a new life at some previously uninhabited place. Particularly on islands, the phenotype of a species can change as rapidly as if a new history of the Earth was about to start on this tiny patch of land. Endemic species arise with high probability — life forms who are at home in only one tiny spot on Earth. To generate this quick explosion of newness, researchers know today, the impact is much greater if a tiny part of the population is isolated, rather than allowing for total competition among all its members. A chance mutation among the few "Robinson Crusoe" individuals then can have major consequences, which without prior isolation would have been completely cancelled out among the bulk of the population.

A founding population is caused not only by a literal shipwreck and rescue on an isolated patch of land in the ocean. An "island" can be anything: a landslide can separate one part of a valley from the other. A tiny mutation can lead to the incompatibility of some individuals' sperm with most of the females' eggs. Rain washing away the seeds of an orchid only for a few yards in a tropical rainforest can transport them into alien territory. Particularly in tropical forests the species number is extremely high, but the frequency with which they appear is rather low — a self-reinforcing diversity feedback loop. In the few isolated individuals of every species in the rainforest, slight variations can have huge impacts: the idler wheel of biological imagination starts to fly free.

We all carry the potential of those tiny, creative mistakes with us. We all can become founders at the right moment. A host of divergent genes describing the possible space for new options can be called to action at any time. With this, we can observe a natural mechanism that continuously extracts newness from biological background noise and that can do completely without the ideas of fitness, adaptation and pressure to perform. On the contrary, the founder effect builds on sleeping potential, not on the overfulfillment of the standard.

What should an objective measure for the concept of selection be anyway? To what standard can it be held? What is the frame of reference according to which an organism can be seen as fit or unfit? Positing such a standard tacitly follows the idea of an objective reality to which a species can adapt in an optimal way. But reality is changed continuously by the species itself. Which life scenario of the millions of diverse forms should be regarded as more apt, more highly evolved, more real? The evolutionary tree is a metaphor, which even mainstream science has given up (although not school education; that will take another 50 years). There is no clear up and down, and there is no objective better or worse. By adapting to an ecosystem the adaptive requirements themselves change.

In the end, a rather weak relationship remains, which has much more similarity with an echo or a resonance or a mutual expressiveness than a biological niche that could be defined in a non-contradictory way. The idea of niche always makes us think of some already arranged, tiny and highly efficient room waiting for its inhabitant to arrive. But in ecosystems there is only opportunity and imagination, not a choice of fixed jobs. Superficially, there might be some recurring roles, such as "eater of hard seeds," which probably in any case would have a fairly strong beak or fast-growing incisors, vole-style. But ecological reality expresses itself on many deeper levels. The degree of entanglement with others invariably changes because the species itself is an active force. Also, here we can see that classical biology has by far underestimated the active role of biological agents.

There is no objective environment in nature. There is only a host of organisms that follow the desire to further exist. They do not only adapt to a preexisting situation but also create novelty. According to the ecologist Michael Rosenzweig, "Habitats are not definite. They only develop through the coevolution of species."[18] With enough time, he continues, species diversity can increase endlessly. It is not dependent on any external factor but is a self-referential process. Seen from this angle, species diversity and the coming into being of new life forms are ways by which the biological possible converses with itself exploring the infinite possibilities of life for which only mere existence is required. Everything else follows in a self-creative manner. We thus still live "in the world's infancy," as French philosopher Bruno Latour once observed.[19]

LIVING BEINGS CREATE REALITY

Remember the squirrels that were seeding pine trees in my garden? Or think of the beavers — engineers creating new water bodies and hence producing entire ecosystems. Organisms always have been a geological force. Organisms create possibilities for others through their own lives, through doing what they need to do in order to unfold to a deeper complexity. Blue-green algae, which were dominant in the oceans billions of years ago, created our current atmosphere through their massive production of oxygen. The open ocean is a particularly perfect example of systems that cannot be disjoined into stage and actors. In the empty space of the sea the only available surfaces are the bodies of other beings. Therefore, animals settling here do not push away others from their space but rather attract them as a nightly lantern attracts swarms of moths. Drifting colonies of gelatinous salps are populated with small crustaceans and worms and are surrounded by smaller and bigger fish. Coral reefs in tropical seas — but also the clam-flats of northern tidal coasts — are biotopes whose possibilities expand rather than shrink with every inhabitant who arrives. Every participant in the round dance of life raises the number of possible dance figures. By this process, during the multimillion-year history of coastal reefs, an ever-growing complexity built up, ever more layers of interactions were welded together and ever more niches for different ways of beings were established. As in tropical forests, researchers cannot explain this diversity in a satisfactory manner by using their selections models alone.

One thing, however, seems to be highly important for self-organizing creativity to unfold — that relations between even the tiniest inhabitants of these treasure groves of being are not based on the generic distribution of the roles of predator and prey, hunter and chased. Every relationship has its individual narrative. It has been molded into a highly specific form by the mutual history of two or more lifelines. In the open waters off the ocean coasts, for instance, many species of tiny herbivores — microscopic crustaceans, minute shrimp — have relied on a narrow spectrum of food types. They consume their particular prey species, but they are totally dependent on them at the same time. Can we still call such coevolution an arms' race? "Algae in the course of evolution develop better and better shells, herbivores more and more specific mouth structures to crack these armours," observes marine biologist Victor Smetacek of the Bremen-based Alfred Wegener Research Institution.[20] To feed efficiently on a certain algae species means to be unable to eat anything else. Both sides of this relationship are inextricably tied to one another. They have tuned in perfectly to their respective traits. Certain crustaceans can only feed on specific algae. When these perish, their hunters disappear. In the oceans, therefore, a struggle for life takes place in which it is only possible to win or lose together. It nearly seems as if hunter and hunted were different facets of one and the same structure, mirroring one another, imagining one another. Only when researchers start to rupture the web of life in order to isolate one species and to concentrate on its traits alone does the species start to acquire the traits of predator or prey, egoist or compliant victim. But in reality there is only the whole, through which these characters emerge in a fluent and changing way, as aspects of one being, not as individual roles.

Therefore, we must accept that if we miss even one of the silver threads from which the web of life is composed, we lose a part of our own freedom. A world without seahorses or lugworms or nightly moths would be a world in which we are infinitely poorer. We would lack an infinite number of our own possibilities. With every vanished species we lose a serious amount of freedom. We need these options, this freedom, in order to retrieve all the possibilities of being alive, which are hidden in our own existence as embodied beings. Why are city kids still mesmerized in the face of lobster, a fish, a dog or a lion? Why, if not because of the fact that animal experience pervades their souls, because the grace, the sheer pleasure in existing that other beings express, is an illustration of their own options for being, inwardness having become form, an inside in an outside, and only graspable as such.

The sea unlocks our view. Its colors, its unexpected transparency, opening like a lens to the floor of an otherwise hidden world, is a magical mirror in which we not only admire the versatility of life but — entangled with all their aspects — always also see ourselves. The ocean floor that opens up during low tide is the unconscious dimension of organic being including our own. We walk over it, and we recognize its beauty with bewilderment. The delicate water layer on the immaculate sand reflects the sky and the clouds and the dark rocks in silvery and violet hues. The clouds and the foam crests of the approaching waves blur into a landscape that reveals itself by being nothing but a gesture, wordless, speechless. The world has converted into one's own soul. Our path follows the shimmering edge where the surf thoughtlessly caresses the earth.

Part Five:
Symbioses

On the following pages I will describe the extent to which we are not separate from, but coexistent with the biosphere and why we only can understand ourselves from that viewpoint, as single agents entangled with a vast network. We are living in symbiosis with an ecosystem whose depth and extension we cannot grasp and which is continuously regenerating itself and us. This symbiotic relationship is material as well as mental. The other beings incorporated in this colossal web, therefore, are neither only metaphorical expressions of our own existence nor others we need in order to thrive and to express our emotional identity. They are literally part of us. They form parts of our bodies, and we participate in theirs. We are part of their metabolism, and they partake of ours. We *are* the living world. We are connected to it in a way we cannot untangle. No human is imaginable without his symbionts. He is a subject, but this subject reveals itself as a bundle of multiple subjects comparable to a whole biotope through which stream the torrents of the world without any guiding agency in control. We are subjects without a firm center. This also applies to our personal identity. Our ego is not a fixed point. It only arises through interaction with the world, and finally dissolves in it again.

CHAPTER 10:

The Body of the Sea

*Water carries us. Water rocks us. Water puts us
to sleep. Water gives us back our mother.*
— Gaston Bachelard

When I gazed down into the boat from the jetty, it looked minuscule. It seemed not much bigger than a toy craft kids play with in a wading pool, gleefully sweeping it across the surface then sinking it again and again. The tiny red plastic hull jerked irregularly in the choppy waves rolling in from the Atlantic. The boat was tied to the jetty with a cord knotted to a rusty metal ring in the stonewall. On its bottom a puddle of murky water swayed back and forth, generating its own miniature swell. Above it, at the top of some rusty rungs in the jetty wall, stood the guy I had already observed on the beach over the last few days. I had thought he was one of those German dropouts who haunt the Canary Islands, hanging in the mild climate, doing nothing specific and spending as few euros as possible. Many of them seem homeless, as people in a benign climate can be, faces deeply tanned, long hair baked together into dreadlocks, always in shorts and thongs. I was baffled. I had not expected that someone like this would carry me out to the open Atlantic.

I could read the name of the boat on its bow: *Canope*. The humble vessel had been named after the ancient Egyptian city near the Nile delta, which gave its name to the canopic jars containing the conserved intestines of a mummy. The manner in which the young skipper pronounced the name, however, had nothing mythological or even sophisticated about it; it rather sounded like ca-nope. Or canoe. And indeed, the tub that I unknowingly had booked yesterday for a whale-watching tour off the Canary Island of La Gomera, fitted the mispronunciation quite convincingly. "Hi, I am Christian," he said to me with a tiresome voice. One of his feet — both of which were adorned with cracked Adidas slippers — was wrapped in a dirty bandage. We were waiting for the other passengers. I climbed down into the boat and let the swell rock me, the long waves that ran out in the small harbor of the coastal town of Vueltas. The sky was gray. Unlike the other days, when this stretch of coast had remained in the lee side of the howling trade wind, today the gale surged directly into the bay. It made the pennants and the rigging of the yachts flap and ring nervously.

"Could one of you guys work the pump?" Christian asked, as the *Canope* lurched over the first huge waves outside the harbor. Water pattered down into the hull and deepened the pond accumulating at the bottom. I was not to be asked twice. I started to operate the rusty handle protruding from the already submerged floor planks of the vessel. Two of the other passengers cheered with pleasure whenever the bow hit a wave sending spray all over us. One of them, obviously a friend of Christian, a guy in a torn T-shirt, hunkered in the bow, clasping a two-liter bottle of beer. At his side, a middle-aged woman in a long flowing gown exposed her face directly to the sea spray, her eyes closed in reverie. On the middle bench sat a third fellow traveler, a retired police officer from Bielefeld, as I later learned. Like me, he'd quickly put on a life vest as we left the harbor and steered towards what, from our humble view, appeared to be a row of gray walls made of water — the open Atlantic.

From nowhere else off the European coast can you observe as many different whale species as off the Canaries. The islands, though politically part of Spain, belong more to Africa than Europe. The shores of these volcanic islands brusquely drop down over a mile to the mid-Atlantic ocean floor. Close to the mountainous islands, species are to be seen that usually shy away from the shallow ocean stretches that normally surround the bigger landmasses, including specimens of rare lineages that almost no one had ever caught sight of until now. Only a few days ago, the rumor went, some tourists in a tiny boat like mine had spotted a blue whale, one of the last heirs to the title of the biggest of all mammals. The Canaries are a truly advanced observation post. They are encircled by the routes some cetaceans travel from the Northern to the Southern Hemisphere during the course of a year. When I decided to come here, I knew with certainty that I would at least see a couple of dolphin species, plus surely some pilot whales, of which a huge population thrives off Tenerife. And maybe I could spot one of the rarely seen gigantic ones.

Marine mammals had never attracted me with the magical force they exert on some people. I did not want to encounter them for any esoteric reasons, or so I thought. I did not look for the mystical awakening that drives a whole industry on La Gomera, a mean-ing-of-life-industry, which is mainly run by German expats. They have founded businesses like "DolphinSolutions," with services inte-grating "dolphin therapy, NLP, shamanistic, spiritual and ritual work and dolphinic-systemic constellations." I had not flown here looking for any mystical answer to my life's problems, at least not consciously. Nor did I seek any spiritual revelation. I rather wanted to understand the web of life from a scientific point of view. Still, I was looking for the kind of science that would not omit my own feelings. I thought that an explanation of the real world that pushes aside a central part of that reality — and particularly that part of reality that I knew best, namely personally meaningful experience — could not be an all-encompassing explanation. And it could be no model in which to whole-heartedly place one's trust, let alone a foundation to build the future of humankind upon.

Whales for me seemed to represent some other unspoken aspect — not a mystical healing consciousness, but the fact that there is a subjective and meaningful dimension to that which is normally re-garded as the area of plain biology. We can connect with them, not because of some brain capacities we share with them, but for the simple reason that like all creatures they are sensitive bodies being and becoming in their world. Indeed, whales in particular, for me, seemed an extraordinary way the oceans expressed themselves, a manner in which the aquatic character of the planet became mani-fest and subject to experience. I knew somehow that I had to get to know the whales if I truly wanted to understand what linked me to all the other life forms. I followed in the wake of the idea of a universal symbiosis. But I knew that this did not mean an understanding of all life as total harmony. I was here because I wanted to learn the price to be paid in order to be, and stay, in connection.

In the last chapters we have seen that viewing everything in biology in the light of utility is not sufficient to explain the diversity of forms and appearances. Here I want to explore why it is equally not enough to focus on one or a few species alone in order to understand the enigma of life. Always, like a unified fabric, the whole texture is concerned. You cannot isolate single strands. If you touch one thread, the entire interwoven network stirs. If you pick it out, the web tears.

The sea can teach us this truth in its purest form because we experience its waters as one solid body from the outside. The ocean. But in reality this one compact body is a fluid mix of uncountable singular figures down to the most basic molecular level. When a whale breaches the ocean surface to breathe, for some fragments of a second we can witness the transition from the whole to its part. We watch the concretion of a hidden potential. The water divides, sets free an individual for some moments and then takes its back into itself. When the whale blows its fountain of spume, the ocean bends upwards like a steaming breaker. In the next moment though, this wave converts into a massive individual body. That which has no form takes shape; or rather, we see the proof that the amorphous mass of the ocean is composed of living individuals in the same way a dense solution is composed of molecules. If you take out all the living particles, no single drop of "just water" remains.

The ocean teaches me again and again that in order to understand one being, all are of importance. No doubt, the biosphere of our blue planet is one single interwoven system. In some respect, its individuals can be likened to the foam crests upon the ocean's waves. They are themselves, and yet they are also the whole. Symbiosis, I want to argue in the following pages, is a phenomenon that marks every organism from the most simple functions of the cell onwards. If you want to understand how the molecules of a single cell cohere to grant a continuous life process, you have to put symbiosis in your main focus. Ecosystems can be read as general competition, but they can as well be seen as a universal, cooperative codependency. Symbiosis is everywhere — and it is visible in all places. This incredibly complex interconnectedness can be witnessed in the organisms' bodies, in their gestures and in their behaviors. Every living being is a monadic mirror of the whole that becomes transparent through it.

In this and the following chapters of this section I want to describe the degree to which a symbiotic viewpoint defines poetic ecology's perspective on the world. The journeys I will embark upon in the following chapters will lead us quite far away from our own assured individuality — up to the final question we must ask ourselves, namely, *are there any individual beings at all*, or is the biosphere to be understood as a nexus of *intersubjects*, in which every single one of us participates. With this, I approach a standpoint that goes beyond the findings reached in Part Three, where we looked at the processes by which self and other interrelate to form a fleeting subjective identity. Other organisms are not only necessary to understand ourselves, but we need to be in a physical connection with them in order to exist in the first place. We need to be many in order to be one. We therefore have to explore the fact that all subjects in the biosphere are far less distinct from one another than we think. Rather, they might be interdependent manifestations of one superindividual whole — neither a self nor an other, but a living body that we can feel as our own.

Such a standpoint might seem rather abstract. But it is not. It is, on the contrary, exactly the idea of radical symbiosis, which could help clarify a host of open questions in biology and open up new interesting prospects, from the sudden appearance of new species to the coherence of ecosystems, from our body's unity to the constancy of our thoughts. This section, therefore, is dedicated less to the individual and the laws governing its expression than to the whole from which the single organism unfolds as a flower buds from a plant.

Already on the crossing from Tenerife, where the airplane had landed, I had been deeply immersed in this experience of intersubjectivity. I had taken the small jetfoil ferry to La Gomera. It whooshed over the long silvery Atlantic swell, filling me with excitement and gratitude. On the stretch of open water between the two islands we encountered a school of pilot whales, whose massive bodies emerged from beneath the rugged ocean surface in low arches, through a glistening curtain of water droplets. In jet black, gray and shining silver, the huge animals seemed to be poised in the sparkling air above the water for some moments before they cut into the next wave again. While our ferry pitched heavily against the swell, the whales moved graciously as if they had no weight at all. They seemed like the perfect product of the element in which they swam. Every one of their massive, and at the same time seemingly insubstantial, shapes was a muscle in the tissue of a nameless body, which extended in every direction without end.

Now, two days later, the ocean's huge body behaved rather unsteadily. I was not even sure if our small boat had the clearance to leave the harbor at all with such a high sea. There was no storm, just the dynamic surge of a strong trade wind, but in the valleys between the gray waves I felt like I might have been boating around Cape Horn. Christian kneeled in the middle of the tub close to a raw bench that contained the petrol gauge that had stopped working, the radio and the accelerator — a stable base which he clutched whenever a sudden jolt pushed the boat. He needed to jam the tiller to the side planks in a sharp angle with his foot to maintain course. "Her propeller has some twist," he dismissively shouted through the wind, shrugging when he saw my enquiring gaze. In truth, he had just passed the exam for his boat license for coastal waters three months before in the German sea town of Bremerhaven. Sudden gusts of wind made the gray cheeks of the waves shiver. The sea came from athwart, and every time the small vessel started to climb up the next crest the gunwale tilted until it touched the water.

The radio crackled. The crew of one of the other bigger boats out there, a sailing yacht, had spotted the blow of a big whale farther out. Christian confirmed via radio and then pounded the tiller harder with his foot. The *Canope* veered a little more into the wind, which now blew directly into our face. "Are you sure that it is safe to go out there with this boat?" I asked the newly promoted skipper. "No problem," he said but could not finish his sentence because a jet of salty water filled his mouth. The police inspector's face seemed greenish while he stared to the horizon whose outline disappeared again and again behind the gray wave crests invading us. The beer drinker in the bow was soaked as if he had taken a bath. A shearwater was dragged along by the gale in a swift flash, the body and the overlong wings of this open ocean bird tensely stretched out like the bars of an elegant cross.

The wave crests overran the tailboard and gushed over my fingers that clutched the gunwale. The water made tiny white bubbles. I felt as if I was already part of the sea, part of its salty infinity, which hit my face with every gust of wind and surrounded me as an absolute landscape made from gray mountains, gray clouds, gray birds. I sensed how something inside of me loosened, how a part of me more and more desperately clutched the slippery plastic of the gunwale, while another part let go a little more with every jolt, as if all this indeed were a part of my home and this torture only one of its still unknown mysteries.

Then the dolphins came. We recognized them only when they showed up directly at the side of the boat. The foaming wave crests had hidden their arrival, their dull slate color concealing the smooth gray of the dolphins' bodies. We saw them when they shot from the ridge of a wave, first three, then four in perfect sync, leaping several yards above our heads. Their rigid fusiform bodies separated the gray water, sent out a dense spray of flying drops and then cut back under the surface as if the ocean did not pose any resistance to them, as if the sea acquiesced to every one of their movements.

First the dolphins showed up on the port, then on the starboard side of the boat, their leaps always accompanied by a dull puff when they blew out used air like a swimmer in a competition. There were many more animals. Christian let the motor chug more slowly. The waves tossed our ungainly vessel brusquely into the air as if they enjoyed the sudden cut of power. "Have you seen something?" a voice squeaked from the radio. "We have dolphins," Christian replied, and a streak of water rolled down his neck as he crouched over the radio. Then as we swayed high above on the crest of the next breaker, we could see in the valley below dozens of animals now swimming around the hull. Again and again they leapt in groups of three or four as if they were fixed on invisible wheels, which swung them out of a wave at the same exact moment and plunged them back in the next, following lines with perfectly fixed distances in complete sync. In fact they seemed to be a part of the wave and not separate from it. The dolphins seemed like the ocean itself had become alive.

This was not the first time I had been so deeply lost among unknowable other lives, "on the high seas, threatened, at the heart of a royal happiness," as Albert Camus once put it.[1] I was on board a sailboat belonging to the whale research institution Tethys, based in Milan, Italy. I had accompanied a couple of young scientists attempting to map the migration routes of fin whales in the Ligurian Sea off Genoa. The sturdy forty-foot boat had been built in a shipyard near Bremen in Germany in the 1970s. There were some young students from Perugia and Assisi with us. Most of them had become desperately seasick when the vessel left the harbor of San Remo, crossing the course of an Enterprise-class aircraft carrier en route to the bustling Riviera town. But that sea was by far nowhere near as high as this one here in the Canaries. And, thankfully, the chop had calmed down considerably by the first two days, so much so that the scientists decided to stay out in the open even during the nights.

In the evenings, the skipper switched off the pounding diesel engine and let the vessel drift with the swell. We sat tightly packed on the bench astern, which formed a half circle, ate bruschetta freshly prepared in the minuscule kitchen below, drank Frascati from the supermarket and tried to fish in the glow of a strong electric lamp. Its light cone cut a window into the water. It had the form of a pyramid turned upside down, its fringes blurring into the darker and darker outer waters.

Attracted by the light, countless shadows swirled through the illuminated space, like insects clustering around a street lantern in early summer. But these animals were no insects; they were plankton, floating across the seemingly bottomless aquatic space. The sea was filled to the brim with tiny crustaceans, medusae, salps, spiny arrow worms and young fish. The boat-lamp's gleam revealed the ocean's true consistency, which seemed to be that of flesh. We fished with light, and we hooked life itself. In the transparent jar that the lamp had carved out of the otherwise opaque water, the sea took a shape that I had never seen before, one I'd only vaguely guessed. It seemed as if the light cone had started a chemical chain reaction, similar to an indicator fluid, which detects an invisible substance in a glass with clean liquid by transforming it into a bright color. Water was life, which was the result of our alchemical experiment.

During these evenings in the Ligurian Sea, I could hear the waves gurgle beneath the plump hull of the old sailboat. They wrinkled the light cone and made its surface move up and down, come closer for one second and withdraw in the next moment. Behind it, blackness stretched out, from which the huge movements rolled towards us like the unconscious stirrings of an immense body to which we were exposed for better or worse. The boat was suspended in a void, hanging from a delicate turquoise thread of light. I was always astonished at how strongly I had the impression of flying when I was on board a ship. I was part of a swaying and rolling movement that knew no bottom. The day before, when we had taken a bath in the perfect groundless blue of the open Mediterranean, this impression overwhelmed me. There, hovering in the water is true flying because, unlike air, the abyss of the sea does indeed carry us.

Later, when our chatter died away, the sea raised its own voice. The waves faintly sloshed along the bow and rocked the vessel into a rhythmic, spinning trance. The wind whooshed in the cables and ropes, delicately surging and then ebbing away again. A fish splashed through the surface as it thrust into the cloud of creatures in the turquoise shine of the lamp.

And there was another sound, sometimes close, then distant again, irregular, recurring, silent for some minutes and then audible again. It was a vibrant puffing, a dull snort, a bit similar to the heavy breathing of a cow. It was a frightful and unearthly sound and still somehow familiar, as if in the darkness of the rocking water some half-human creature was gasping for air. The whales were blowing. Their breath rushed through the darkness of the Ligurian Sea. It was nearly impossible to spot the source from which their sounds drifted towards us. They seemed to be everywhere, as if their hidden bodies were dancing around the boat in an invisible cone of ubiquitous sound.

Fin whales are the second largest of all marine mammals and in most respects are quite comparable to the much rarer giant blue whales. Only in the last decade has it become established that an independent population of fin whales dwells in the Mediterranean Sea, close to the coastal hills of the Riviera, vistas aflush with villas, gardens and busy towns. At least 4,000 of them gather here in wintertime to feed and reproduce.

In the afternoon we had circled around a young fin whale for hours, trying to capture a tiny sample from his flesh by maneuvering a long spear with a tiny, razor-sharp bucket at its top. Most of the toxic substances a fin whale takes up with his food concentrate in the thick fat layer beneath his skin, the blubber. Before Simone, the chief scientist, finally managed to thrust the spear into its blubber, the whale had dived at least a dozen times. It was late afternoon, the oblique sun giving the ocean surface a metallic shimmer. We had waited for the animal over and over again and in the meantime swayed on a surface of molten silver, tenderly forged to the liquid element as if it was a strange kind of ore.

Despite the promising radio announcement, Christian gave up the search for the big whale when the ex-inspector and I refused to endure it anymore. Still farther out, the waves built up more and more force. When we glided down in their valleys, the horizon appeared so far away that it seemed to never reappear. The wind continued its heavy gusts blowing spray directly in our faces. It seemed to me now as if every jolt could capsize the boat, as if time had granted us a merciful but steadily shrinking respite, until the inevitable. I admit that I did not enjoy my feeling of being lost anymore, not one bit. The impression of having been estranged from my element, having abandoned the "green, gentle, and most docile earth," as Herman Melville put it in *Moby Dick*, filled me with panic.

Reluctantly, Christian accepted our request and turned the boat around. He seemed a bit dull, but one could not say he was not brave. The beer drinker and the woman in her esoteric garment protested. The skipper had to promise he would bring them immediately back into the uproar after having safely deposited us on land. Only when we had reached the leeward side of the promontory did the waves' force begin to lessen. The sun broke forth through the foggy shrouds that had covered the sky and made the gloomy gray ocean at once take up an energetic blue, as if a giant source of light inside had been switched on. The waves came rolling in backwards now, lapping against the stern. Now that they were gentler, I realized that they had a complicated structure with different storeys. Indeed, smaller wind waves camped upon huge oceanic swells. The boat rode upwards on flowing stairs of water and glided downwards over cascades of blue. The waters rolled in on us nearly imperceptibly, forming long and stretched-out mountains with a height difference among them that seemed to be at least as tall as a house. We approached the land suspended over 5,000 feet of liquid blue glass.

At the end of the *Canope's* return voyage we ran into another school of dolphins. It seemed to me as if nothing human could ever replace the endless ease these animals display when they shoot out of the blue Atlantic waters, all in perfect synchrony. Each animal drew a perfect sine curve before it melted back into the aquamarine body of the sea. The dolphins closed up, let themselves be carried by the bow wave and accompanied the boat, now at full speed through the water without stirring a muscle — archetypes of an effortlessness that in this moment I could also somehow feel inside of me.

Christian and the two others that went out again did indeed make it back. But they arrived late in the evening with nearly empty fuel tanks, after the *Canope* battled a strong offshore wind that had suddenly resurged. The three were completely soaked, as though they had been tossed into the sea and forced to right the boat after it had capsized, but they had not; they had only been drenched by the spray of breaking waves. In the harbor bar, having a beer after their terrifying voyage, they pulled dripping money and drenched ID cards from their pockets. They had not seen a single whale.

THE SPIRIT OF SWARMS

The afternoon before, I had gone snorkeling. Underwater, the rocky beach sloped down steeply towards the seafloor already a couple of yards behind the waterline. I quickly slipped into the cool liquid. When I was fully immersed, I lifted my gaze up from the tiny arms of the barnacles under my face, from the sea anemones and algae that swayed in the surge left by the breaking waves. I suddenly had the vast emptiness of the open sea in front if my eyes.

I slowly drifted above the minute details, hovered over all these traces of life whose ways were so different from my own and which were presenting themselves in a strange but perfect order. A joy filled me, which I recognized from boyhood times. It was a specifically marine joy, connected to the iodine smell of the shore, its blue and pink seashells and its promise of meeting life-forms I had never seen before. I remembered this exultation from when I was a kid and I traveled with my family to the Baltic Sea. When I arrived there for the first time, still very young, I instantaneously fell in love with the ocean. It was a kind of immediate, innocent love I am still trying to find words for. It is exclusively linked to the smells of the shore, to the cool substance clutching my body, to the roar of the waves.

When I looked up again, the open sea had darkened. A compact wall had built up and was slowly approaching, splitting into glistening fragments, small ones and others still smaller. The mass, at first heavy and immobile, started to move, almost with the swiftness of a thought. At times it seemed gloomy and threatening, then fragile and fugitive, as if it was smoke spreading underwater. When it came closer I could see that the cloud consisted of fish. They sent their fixed gazes through me as if I was also a transparent drop of seawater. It was a large school of barracudas. Their motionless eyes seemed to be focused on an indiscernible distance. And yet every individual's movements were so perfectly synchronized with its immediate neighbors' that they seemed to travel like one single body.

In front of the shimmering swarm of fish I had the feeling that not only was each single animal alive, but that I was gliding along a living substance. It melted away, joined again and dappled my body with glistening light and shadow. It seemed as if the water itself had turned into a being, revealing its true nature as inspired matter. The swarm was one single silver body, yet many figures, a shape made of countless forms that absorbed me and then moved on, leaving me enraptured. I was poised in the ocean, in front of a cathedral made from living flesh.

Until a few years ago, the number of creatures in the ocean was supposed to be endless. They were simply there, always, and in an incredible abundance. Fish and squid and whales seemed to appear from nowhere like a nameless gift. Nature offered itself with the boundlessness of a miracle. Today, however, the oceans are so heavily overfished that not only are their inhabitants about to disappear, but the idea of gratuitous plenitude itself is vanishing. For to know what real abundance means, we need a reference in the world. To know what the gift of life means we need it to be given to us. The quantities of glistening and winding bodies that poured out of the dark sea give a physical dimension to the idea of plenitude that is nourishing even if we do not work for it. If the oceans empty, these emotions lose their anchor in reality and fade away.

In some remote locations in the oceans can we still experience the glistening of seemingly immeasurable schools of living flesh. Each year in early summer, for instance, streams of sardines move northwards along the South African coast. Each one of these swarms can still be several miles long. Hundreds of millions of individuals pile up and form wandering mountain chains in the water. Their dark ridges cascade down into silvery canyons and then surge up again forming summits made of flexibility and muscular force.

Fish that form swarms, biologists believe, gain a host of advantages. Their individual bodies melt into the metallic armor of the entire group, thus eluding predators. Under this illusory spell the would-be predator aimlessly bites into the wavering silvery blue, instead of finding lunch in a defenseless loner. But this communal defense is not perfect. The carnivores that follow such swarms still thrive on the legions of their victims. New swarms in this way build up from other swarms. The sardines off South Africa graze the plankton herds, but on their account drag cohorts of sharks and dolphins along with them, plus a hurry-scurry of seabirds that thrust with pointed beaks into this display of living meat. And then there are the armadas of factory trawlers.

It is only a question of time until the congestion of the oceans will yield to emptiness. We are ignorantly working towards the abolition of the miracle. Below the waves, the ocean is being changed into an uninhabited mausoleum. In most of the important fishing grounds worldwide, catch has been dwindling for years and more and more often the nets remain empty. The sardines off the American Pacific coast for instance, where coastal towns that landed and processed the catch once thrived, have all but disappeared. Cod, which I remember eating from the preprocessed Alete baby food tins my mother warmed up for me when I was a toddler, is exhausted.

Until a few decades ago fishermen off Norway and Newfoundland referred to cod not in terms of swarms but of mountains. Millions of the huge animals, which can weigh over 300 pounds each, towered along these coasts in dense columns, hundreds of yards high. For months, more and more of these layers of life flushed into the coastal waters to mate and lay eggs. At that time, the arrival of such a mountain range built from sheer energy and determination must have been a breathtaking experience — one of those events that for moments abrogate space and time, and give birth to their own set of categories.

Yet a swarm is not a nameless mass, nor is it one big organism. It is a world made up of individuals. During mating time in this turmoil, males and females pair up for copulation only after they have chosen one another with great care from the confused mass of individuals. Their mating can take a whole day and resembles a complicated and refined dance. Only after long hours of a highly intense courting process does the male cod hover over the back of the female and release its sperm in order to fertilize up to five million eggs, which the female then deposits on the ocean floor. For the lost mountain ranges of the cod, which once enjoyed a formidable conjugal dance, the honeymoon is over.

TOWARDS A SYMBOLIC ECOLOGY OF ABUNDANCE — THE SWARM

Although modern technological power managed to decimate many of those swarms only in the last decades, the big pillage of the oceans was already in motion centuries ago. Scientists, though, have not always been aware of the deep past of destruction. They only recently discovered that what they had believed to be original nature was already a degradation. It was only a faint reflection of the old density that once infused every surface with the glow of life. What is supposed to be the good old times of natural abundance already was a large step back from its original generosity. In the Caribbean, for instance, the Spanish conquistadores had named hundreds of places after sea turtles — coves, capes and islands like Turtle Cove or the Dry Tortugas. Where today a few of those ancient marine reptiles still emerge from the waters in order to lay their eggs, five centuries ago an incredible crush must have been the norm. Several species conglomerated at certain beaches forming colossal *arribadas*. When 50,000 sea turtles buried their eggs on every mile of beach, observers literally could not step on the sand anymore. Ship crews slaughtered the easy-to-catch animals in order to restore their provisions from this seemingly inexhaustible mass — until it was worn out.

The really big swarms are no more. But the void reverberating through the blue vastness of the oceans is more than a broken-down fish market, more also than the loss of a uniqueness belonging totally to others. It is an emptiness in ourselves, a void of the soul. With the plenitude of life disappearing, we lose something of our own inner substance. We lose the dimension that is always beyond any calculation, beyond any ideas of purpose.

The spirit of poetic ecology is the spirit of swarms. It keeps in mind the plenitude that echoes the void, that builds itself up and creates order for free and then piles up silvery mountains made from living flesh, a vast expanse packed to the brim with being.

In Chapter 2 I described the Three Laws of Desire. They related to the single organism. But we have seen during the course of the argument presented in this book that an organism cannot exist alone; it cannot even be conceived of meaningfully in separation from others. Our idea of individuality is hardly the final truth concerning the reality of a living being. Will we need to go so far as to regard individuality as an illusion after all? The experience of swarms shows that at least we must enlarge our scope quite a bit. To understand the individual, we need to understand its environment, and each through the other. We have to think of beings always as interbeings. Therefore, including the observations we have made in the last chapters, we can expand our earlier axioms to encompass ecosystems, or even enlarge them to the whole web of life.

The following synthesis of the principles of a poetic ecology could be a first approximation:

1. Wastefulness is as important as efficiency for ecological unfolding.
2. Psyche is not only inside, but also always externally visible.
3. We are a swarm ourselves, and we form swarms.
4. Mind is not inside an organism but distributed among the bodies of living beings.
5. Individuals are not absolutely separate from each other but form a huge interconnected tissue.

For the idea of a poetic ecology the phenomenon of the swarm is a basic aspect of life. Its interconnectedness is reflected in the barracudas' silvery bodies, in the dolphins' poised ballet, in the tiny fish larvae of the still zones, which I encountered off the Canary Islands. In a swarm life expresses itself as dance. A swarm does not *have* intelligence, it *is* intelligence. The individual's behavior within the group follows simple guidelines. Researchers can simulate swarm behavior in computer programs if they follow the basic rules: "Keep close to your neighbor, but keep distant enough to avoid collisions. Do what your neighbour does." In a swarm, a huge connected whole arises from the local coherence of small parts. This whole, however, reveals more than the simple algorithms that bring forth its behavior. In this respect, any swarm is an intensified counterpart of ourselves. It is what we are and what we try to imagine with our conscious thinking. The swarm does not think. It is a thought process.

The members of a swarm can perform in a sense that exceeds the limitations of the participating individuals exactly because they are not intelligent in the sense of our human intelligence. Their intelligence is distributed, and in order to be distributed, it must have a body, which then becomes expressive of whatever is going on. A swarm does not consist of thinking, reasoning and consciously choosing individuals but rather is assembled from biological magnets, which are attracted and repelled by their environment according to existential principles. Swarms can only show, not analyze. Therefore, the computational explanation of swarm mechanics will miss what the swarm shape reveals. Swarms are solidified feeling. The swarm *is* — and in its *being* living dynamics and their expression are welded together in one single gesture. The swarms' individuals mirror life. But not because the underlying elements follow life's principles purposefully. Rather, they have needs that they need to fulfill if they do not want to risk perishing. The sardines would be lost outside of their silvery cloud. The execution of these existential needs yields the image of the swarm.

Swarms, and together with them the societies of several social animal species, resemble ecosystems on the one hand and whole bodies on the other. Researchers, therefore, call ant or bee states superorganisms. Many of their behaviors can be understood better if we view the whole state as indeed one single being and its members as the cells of one body. Seen from this vantage point, biologists begin to understand that our body can be viewed as a swarm as well, or rather, as a super-swarm consisting of a multitude of lower-level swarms. We consist of billions of single individuals. Our body cells in principle and in the test tube are able to thrive alone. In order to function as a whole organism, they need to tune their behaviors in to one another. In doing this, our cells do not obey an overarching control center. The brain does not guide the coherence of the body but rather is spread out over the whole body by means of the finest ramifications of its nerves and the liquid messenger substances which float through all body tissues. In many respects, therefore, organisms can be understood as ecosystems that drift through a series of states, and in which changes simply come about instead of being centrally managed and controlled. Thoughts, emotions, recollections and plans, therefore, are not contained "in" the brain like record cards in a box. Feelings and ideas are connected to a body and to its swarming cells, which may thrive or die, just like the glistening sardines. The single living elements in each case are afloat in their liquid substrate and act according to simple but existential rules, which bring forth a body, a self and our experienced inwardness. These rules emerge in the core self we encountered in Chapter 7. They follow from the primary emotion of what is good or bad for the individual organism. They are present not as abstract algorithms but concrete feelings.

From here once more we could say that the swarm symbolizes an aspect of the ocean's inwardness. It is the cognition of water. Our emotions for us are "inside," but they still are swarm-like in some respects. They are the expression of a life-process, and as such they cannot be separated from the environment. For this reason they are not crucially different from the existential meanings embodied by the other beings dispersed in air and water, and which consist not only of living cells but of living individuals. We see gestalts of the living that behave according to simple organic laws mirroring the great constellation that every living being has to cope with: to persist, to be close to the other, but not so close as to collide with him. These are the principles of poetic forms that are so thorough we can even teach them to a computer. They are the primary shapes of a poetics of living things.

Swarms reveal what is inside us to a depth of the flesh to which no word can intrude. They show what is unspeakable. Because it is ineffable, it remains tied to living form and, therefore, shares tenaciousness and the immeasurable fragility of all bodies. Every individual is the unique result of the existential experiences of the whole, as the American deep sea explorer and naturalist William Beebe realized over one hundred years ago, when he observed: "The beauty and genius of a work of art may be reconceived, though its first material expression be destroyed; a vanished harmony may yet again inspire the composer, but when the last individual of a race of living things breathes no more, another heaven and another earth must pass before such a one can be again."[2]

CHAPTER II

Not One, But All

No living creature is a single unit.
It is always a multiple.
— Johann Wolfgang von Goethe

Rarely anywhere else is the "Siamese connection between all living beings" (as Herman Melville called it) more salient than in the ocean. He who allows his hand to dangle down from the gunwale of a small boat into the blue water of the ocean can make out by touch alone the primary axiom of ecology. This has been well formulated by the American nature explorer and writer John Muir who observed: "Everything is hitched to everything else," echoing the British poet John Donne's realization that "No man is an island, entire of itself ..." In the ocean, we can feel this entanglement through our skin. The sea water feels soft, even slightly viscous. After drying it forms a palpable lining on the epidermis. Sea water, therefore, is more than an assembly of H^2O molecules with some salts added. In its consistency it is similar to a body fluid. She who moistens her hand with it touches a living universe.

In other times, people living close to the sea had names for this living potential. British fishermen called it "feed." They knew that the slimy ingredient in sea water was necessary to grant them a good catch. The murkier the water, the more herring they could haul in with their nets. At the beginning of the 19th century the French historian and philosopher Jules Michelet imagined that the ocean contained a nourishing primordial substance, a kind of fluid aether of life.

Some years later, on the North Sea island of Helgoland, the German biologist Johannes Müller had the idea of filtering sea water and looking at what it consisted of in reality. Müller used a net with a meshwork finer than that of a lady's refined silk stocking. Müller carefully washed the streaks remaining on the gauze into a glass jar. Then he looked through the microscope and saw — life.

Unprecedented life in a richness of forms that took his breath away. In the bright circle beneath the eyepiece minuscule green cells were poised in the light, some of them with delicate symmetrical openings and regular patterns in their shells, others protected by tiny shining armors. Transparent crustaceans were beating with pinnate body extensions, snails that looked as if they were made from liquid glass sailed along with long wings and slender worms darted through the light. With his simple makeshift silk net, the biologist had stepped through a magical door into the world of the plankton.

The word *plankton* comes from the Greek language and means "that which is adrift." It was invented by Müller's disciple Victor Hensen at the end of the 19th century and refers to all those animals and plants in the sea that are mainly transported by wind, waves, ocean streams and tides. These organisms are often too minuscule to allow for any choice of direction — like all the different larvae of fish, worms, jellyfish and crabs, which are counted among the plankton. Other planktonic animals are huge but do not bother to move, like *Mola mola*, the open ocean sunfish, which drifts like a huge six-foot saucer on the surface and slurps in jellyfish gliding aimlessly along like their predator. The organisms that make up the community of plankton are most distant to ourselves. They are restlessly carried on by the ocean's surges, immersed in the twilight of endless halls of open water, a world without tangible features where nothing solid grants support or orientation and where the only surfaces are provided by the bodies of other beings.

Plankton are everywhere, rocked by the surface waves and hidden in the gloomy deep sea. In some ways the salt water already *is* plankton itself. As if it was a giant organism, the liquid ocean is filled with living organisms. One milliliter can be populated by 200,000 of the tiniest cells. There are (10^{20}) stars in the universe, but many times more planktonic cells in the oceans, as the marine scientist Victor Smetacek of the Alfred Wegener Institute in Bremen calculates. And many of them are extraordinarily small. Take *Proclorococcus*, whose single cell measures just 0.8 micrometers — half as large as an *E. coli* bacterium, and only a tenth of the size of our red blood cells. Scientists assume that *Prochlorococcus* is the most widespread organism on the planet. These countless cells tirelessly transforming light into life are something like the blood cells of the sea or the chloroplasts of the planet.

Through its plankton, the sea sucks up the sun's energy and converts it into countless individual destinies. There are the primary producers, mostly single-celled algae thriving directly on light, and thus providing a substrate in which the bigger forms, which we encounter more easily, can thrive. Medusae with transparent, aquamarine-colored sails drift across the waves of the oceans and drag a veil of deadly stinging tentacles behind them, tiny crabs batter incessantly against the pull of the depth with their pinnate limbs and swarms of glassy shrimps and transparent squid build up opaque layers in the depths, through which a ship's sonar cannot pierce. These migrating strata of animals are so thick that submarines can hide beneath them. The global oceans have an average depth of almost two miles. Everywhere they are shot through with organisms. The seas contain 99 percent of the habitable space of the earth — and this space belongs to the plankton.

Researchers are still discovering the ocean's miracles and beauties, and they are still refuting common prejudices. For a long time scientists thought that the ocean's middle waters were basically empty. When researchers went down in a submersible and looked out of the hatch windows, they stared into a space of black night, a void, like a pause of consciousness, until their vessel had reached the ocean floor. Only when the first scientist thought of switching on the lights of their submersibles during descent did they realize with amazement that they were falling through clouds of living beings.

Many organisms also remained unknown for a long while, because researchers' common catch methods destroyed them. Particularly fragile animals break in the meshwork of a plankton net and cluster together at the bottom as an amorphous goo. Marine researchers regularly found this slime in their nets, but no one really classified it as the organic flotsam that is was, the residue of living beings. Jellyfish are especially sensitive to being squashed in plankton nets. Equally fragile are salps, transparent barrel-shaped animals, which filter algae from the water with enormous efficiency and hence are crucial for the survival of all other members of the oceanic ecosystem.

We owe our recent knowledge about these latter organisms to an American biologist. On a mild summer day in the Gulf of California in the 1970s, Bill Hamner had the sudden impulse to jump into the water surrounding the research vessel upon which he was traveling. Hamner had no special research interest in mind; he just wanted to kill some spare time while his colleagues were occupied by some banal, but time-consuming experiments. So he put on a mask and plunged under the surface of the Gulf which was some several hundred yards deep. When Hamner's eyes accommodated to the gloomy light, he realized he was not alone. Everywhere around him, in what had seemed like blue emptiness, floated the transparent bodies of salps, medusae and comb jellies. They formed a transparent glass menagerie as if the water itself had taken on shape.

These elusive beings cannot be trapped with nets. Salps are inevitably destroyed by traditional hauling methods. If you want to understand their biology, you must swim with them. And here, immersed in the bottomless blue, an unknown world opens up. Researchers who have glided down into this realm carrying sample jars and cameras have met floating snails that wove yard-long fishnets made from mucous threads into the silent blue. When the researchers came close, the snails whizzed away using their transparent wings, a bit like hummingbirds intending not to be caught. Amazed, biologists observed how comb jellies, propelled by long rows of undulating iridescent bristles, single-mindedly traveled to higher water layers to encircle their prey. Nary a biologist had suspected such tactics of an animal as small as a cherry and not even owning a true brain. In a way, therefore, with all these busy beings, where before humans had imagined only void, it not only seemed that the water had revealed itself as actually consisting of transparent bodies — it is also knit together by a host of individual intentions.

Plankton, therefore, is not the formless soup of chlorophyll and protein, of apathetic cells and mechanic grazers, which generations of biologists had more or less dismissed it as. It is rather a microcosm filled with an abundance of willful agents. Only everything is so much smaller; a whole ecosystem can be restricted to a water droplet. The right scale to study plankton is one microliter, says the biologist Farooq Azam, who works at the Scripps Institution of Oceanography in California.[1] One microliter is a thousandth of a milliliter. In such a tiny volume, Azam claims, one huge single-celled algae can share its life in symbiosis with thousands of bacteria. These feed on the sugar the algae excretes and in exchange shed phosphate and nitrate, which the algae use. This tiny ecosystem is a closed loop that only needs carbon dioxide and light. Seen from a different angle, the whole ecological circle in a water droplet seems like a particular version of a single cellular unit. It is a closed ecosystem, a stepping stone between cells and biotopes.

If we can understand tiny ecosystems like the microbial loop of the oceans as similar to organisms, would it not be equally valid to regard the inhabitants of the oceans living in a constant interchange with their liquid substrate less as single organisms and more as units in one huge living system? At least in the realm of plankton with its multifarious dependencies and intricate feedback loops, what a single individual is becomes blurred. But in this respect the miraculous open sea might not be the exception to the rest of the biosphere but rather the rule. If we look at ourselves (and the inner ocean of our physiological milieu, which we embrace inside of us), we have to concede that we are also multitudes. We are symbiotic systems which consist of a diversity of different species without which we could not survive, like the gut bacteria that we need in order to take up nutrients. We also consist of microbial loops.

"NO LIVING BEING IS A SINGLE UNIT"

The fact that organisms reveal themselves to be miniature ecosystems and that members of ecosystems on the other hand can have such strong ties among one another that these associations must be viewed as new single organisms — this has been a fact since the beginning of the history of life. All higher cells, including our own, have only come into being through the fusion of smaller unicellular beings. They were born when cellular precursors swallowed others, but instead of digesting them, incorporated the alien cells within their own bodies as useful helpers. This view, proposed by the endosymbiotic theory, claims that the main gear of the modern cell came into being through symbiosis. If we look carefully, the theory says, we can still see that each cell is a microscopic ecosystem that binds together diverse players. The mitochondria, the energy power plants of the cell, go back to the proteobacteria, and the chloroplasts that provide for the energy extracted from light descend from blue-green algae, which once a long time ago had been eaten but not digested. The process was helpful in both directions: the newly ingested cells could freely use the metabolic products of their hosts and were protected against attacks from the outside.

The endosymbiotic theory was a daring claim when the American evolutionary biologist Lynn Margulis picked it up from the Russian botanist Konstantin Mereschkowski and reformulated it for a broader scientific audience.[2] I remember that when I was doing my undergraduate studies in Freiburg in the 1990s I had to study the then still competing compartmentation hypothesis. It claimed that the cellular subunits had come into being through involutions, folds of the outer cell membrane, which had then detached inside the cell and thus created distinct spaces. Indeed, even 20 years ago it was not yet decided which approach would win. After all, the endosymbiotic theory was a bit outlandish because it bravely blurred the distinction between different individuals, different species and between individuals and ecosystems. Today, however, it has been widely accepted as the best explanation for the ways our cells evolved. Lynn Margulis's reputation changed from dubious rebel to distinguished scholar. But although mainstream biology has swallowed the endosymbiotic theory whole, it still has not really digested its consequences, which are that symbiosis is at least as important a factor of biological creativity as is competition. The situation reminds me of the fact we have discussed before, namely, that genetics introduced the categories of code and meaning but still holds itself to be a mechanistic science. To put the symbiosis at center stage of the history of organisms should, therefore, inspire us to move away from thinking in terms of individual competition. The key to understanding an organism might hide in the paradoxical fact that it is at once one and a multitude. At first sight, this does not make understanding easier, but it allows us to be honest about what there is. It shows a track on which many amazing findings are still waiting to be revealed.

By now we can hardly overlook the proofs that all multicellular organisms are intimately entangled, despite the diversity of species within the tree of living beings. Scientists are discovering more and more patterns that can have arisen only through an intimate joining of bodies and of forces. Probably not only mitochondria and chloroplasts were autonomous life forms in their original state. It seems that also the flagellum, the "motor" of the cell, the spinning whip at their end that speeds the bodies of particular cells forward (as typically seen in male semen) was originally another bacterial species. The underlying thread-like structure in modern cells comes in a huge variety of functions, from propellers to sense hairs. The forebears of the outer body of our cells might have been early forms of the primeval Archaebacteria. Many other tools inside the cell could stem from bacteria from the group of spirochetes to which belongs the bugs causing syphilis and Lyme disease. Here we can also see that a clear distinction between "functional friend" or "foe" does not make sense in biology. The spirochete heritage particularly has led to a blooming of biological imagination within our bodies. These endosymbionts presumably converted into the bristles lining the inner ear, providing for our proprioceptive senses through which we calibrate our body balance and into the cones and rods of the retina that give us vision. Without spirochetes, therefore, our eye as it is today would not have been possible. That means, paradoxically, that without the intimate affiliation with other lives inside our bodies, it would not have been possible to see a distinct external world outside of us. Individual and background ecosystem, inside and outside, competition and cooperation — they are fused and interwoven, and one is always the shadow of the other.

"Particularly man is not homogeneous, but assembled," Lynn Margulis observes. "Each one of us presents a gorgeous environment for bacteria, fungi, roundworms, mites and others which live within us."[3] Some 250 of our 30,000 genes probably derived directly from bacteria. But this number pales with regard to the role viruses play in making up our genetic identity: at least one fifth of our genes are of viral origin. They stem from pathogens that have inserted themselves into our DNA and thus have converted from parasites to parts of the organism. Ten percent of the dry weight of the human body is bacterial body mass. The number of microbe cells in our bodies surpasses the amount of our "own" body cells tenfold, and the number of microbe genes is 100 times as high as our own.[4] And all this diversity is "us" in some respects. Only by the assistance of our symbionts can we process many essential nutrients, for instance, diverse sugars. Some ethnic groups in Papua, New Guinea even carry symbiotic nitrogen-fixing bacteria in their guts, reminiscent of the way many plants host these symbionts in their roots. These people can support a pure vegetarian diet for years without showing nutrient deficiencies.

Partners in a symbiosis remain autonomous to a certain degree. They do not simply convert into functional tools of the host organism. The host does not control them but maintains a balance between all participants; that in the end all together *are* the body. If we take seriously the role taht the autonomy of so many individuals plays in symbiosis, and if we add the fact that hardly any organisms are imaginable without symbiosis, then it becomes strikingly apparent that organisms are ecosystems made from many species and are hitched to their environment in diverse ways. We are not able to sharply discern a biological individual from the background made of other beings. Accordingly, Lynn Margulis and other biologists believe that symbiogenesis, the history of how symbioses establish themselves, is vastly underestimated as an evolutionary force. "Not one organism visible with the naked eye has evolved from one single ancestor," Margulis says. She goes one step further: "Every organism that can be assigned to a particular species, that which thus possesses a genetic identity, results from symbiosis."[5]

Individual bacteria do not have fixed genotypes as "higher" organisms have. They are able to exchange genes in a way resembling the manner by which other organisms take up food. By means of minuscule tubes individual bacterial cells can funnel parts of their DNA into the individuals of other bacterial species. DNA is highly flexible among them. Margulis thinks that the core principle of modern evolutionary theory, evolution through mutation, is only applicable to bacteria because they alone are not assembled from a host of different cells. But the sharing of DNA information could also be seen as a kind of symbiotic interrelatedness. The history of cells, Margulis holds, can only be understood if we take into account their birth by symbiosis. The identity that characterizes higher cells is achieved only by means of cooperation, through integration of the other. For Margulis the biosphere has to be understood less as a war scene of continuous conflict than as an interwoven pattern of structures, needs and dependencies. Her view of symbiosis does not exclude conflict; to the contrary, it explains life history through mutual transformation. One species might exploit another, but at the same time it can be entirely dependent on its presence. We can witness a taking that does not work without a giving, a continuous reciprocation.

NO PLANTS WITHOUT SYMBIOSIS?

One realm of beings has chosen symbiosis as its explicit functioning principle. This is the lichens. For a long time nature researchers were even not sure how to classify these amorphic objects that grow everywhere on stones, tree trunks and meager soils. Were these flat crusts and lobes outgrowths of plants? Or mineral efflorescence? Moss in an atypical shape? A little over a century ago botanists proved that in a lichen a fungus and an algae are cohabiting. Every lichen species is really two species and only thus becomes able to perform in ways that are inaccessible to their conspecifics that live alone.

This symbiosis between sun-eating green plants and detritus-feeding fungi allows lichens to endure extreme conditions surpassing anything organisms were thought able to endure. Lichens can be freeze-dried or pulled through boiling water without decomposing. They dwell in deserts and in the summit zone of the Himalayas. And all because they have developed a sophisticated form of symbiosis. Lichens can thrive where soil has not settled yet, where there is no support for delicate plant roots. They are pioneer colonizers on volcanic islands and among rock fields. Lichens were the first organisms to continuously settle on Surtsey Island in the 1960s after the volcanic patch of land had appeared, steaming forth from the waters off Iceland. Some researchers think it unlikely that vascular plants made the first step out of water and settled on dry land, but instead, that this was accomplished by algae in alliance with fungi. Indeed, it is imaginable that the only organisms with a chance to persist would be those that were able to cling to the bare rock and could survive extended periods of drought.

Yet what lichens do is no exception in botany but rather the norm. Everywhere plants cooperate with mushrooms. Very few green organisms can do without the tiny fungal network in the soil. Over 95 percent of all plants need to be connected to its ubiquitous filaments and threads in order to thrive. Their roots are linked to the mycorrhiza, a delicate meshwork in the ground, complete with slender threads and tiny hubs from which the plant gains access to minerals.

In turn, the fungi are fed with sugar. Only this mutuality gives the plants their ability to survive harsh conditions. Researchers have proven the power of specific plant-fungus alliances by experiment. They conveyed the symbiotic soil partners of heat-resistant grasses to the roots of highly sensitive melon sprouts, enabling them to resist temperatures of over 70°C instead of starting to wilt at 38°. It is imaginable, therefore, that even trees are not the authoritative and solitary giants we have come to see them as, but rather the efflorescence of a network of mutualities, parts of a whole that would not be possible without the existence of all of its parts.

WE NEED THE OTHERS TO THINK AND BREATHE

In the face of ubiquitous symbiosis we can perceive life on our planet in an entirely different way. We do not need to see it any longer under the perspective of isolated participants in a capitalist free market, each fighting for their individual happiness and only being able to thrive if there is competition. We can now view the biosphere in terms of an integrated, interdependent system whose parts are so intimately entangled that neither the contributions nor the needs of one single participant can be distinguished from those of all the others. In such a system any unentanglement amounts to amputation. The other always and inescapably is part of the self. It is required for the self's ability to exist.

So again we run into the fact that we have encountered many times during this book and that deeply characterizes the idea of a new biology: there is no individual sovereignty apart from the totality of life. For this reason, every organism in some respect already represents this totality. It is coupled to everything through its need, which makes the other a necessity. Any idea that posits a concept of complete independent individual autonomy as possible or even desirable results from a deep illusion about what life is. It is true that our understanding for many centuries has been distorted by exactly this individualistic vision, the Enlightenment dream of a free agent commanding matter to which he is not connected at all. But connection and mutual transformation is all there is in living reality.

Mutuality already characterizes metabolism, the ground zero of biological existence, as we have seen in Chapter 3. Metabolism is a process by which an individual transforms that which it is not into itself. It is a process by which world becomes self, and by which self, in its existential expressiveness, becomes the symbol of the world. Can we conceive a more intimate connection of self and other? The same existential mutuality-as-reciprocal-transformation also characterizes the development of psychological identity, which is dependent on the other to grasp itself, like the Ethiopian wolf's black eyes are needed to disclose the bottomless vigor inside oneself. Indeed the still unshaped ego of the newborn child is dependent on the loving aliveness in the caretaker's eyes, which enables her to understand the world in terms of the map provided by her own body. Nobody is only herself. For this reason, evolutionary egoism, as we have been conceiving it, cannot be the deciding driving force alone. There is no such ego, whether we look for it in the species, in the individual or in the genes. There is only one huge fabric in which everything depends on everything and where all move through the movement of all. Here, who could be the "winner" in the "struggle for life" anyway?

Mutuality, therefore, is the principle of the individual body as well as the law governing the interplay of all bodies. And it is the key to understanding reality. For the newborn opens up a cognitive access to her own body and to the corresponding inside, which manifests as emotions only through the expressive body of the other. Only mirrored in the other can the swell and roar of the self's inwardness acquire an emotional form. It seems that living beings, through this process of coming-into-being, behave according to the idea of the Russian psychologist and philosopher S. L. Rubinstein, who thought that a "mirroring reflectiveness" is the prime characteristic of reality. From this vantage point, symbiosis seems the principle of bodies and of mind.

All life is meshwork. All life is relation and transfiguration, metamorphosis through connection. The veins of a leaf, the branching tree's crown, the capillaries under the skin in an infant's hand — if you watch carefully, everything alive reveals itself to be a pattern of connection and superposition. All living processes are tissue and swarm. If you look carefully, what you consider to be your "self" emerges as part of this existential connective tissue. If we could for some minutes expose a photographic plate to the movements of the planktonic crustaceans in the turquoise beam of the lamp, then the tracks of the tiny animals would resemble growth paths with which plant roots interlace the soil. Indeed, both would bear a striking similarity to the entangled cluster of lines in a schematic diagram representing a marine food web.

In biology, nothing is to be analyzed separately. Everything holds true only according to the whole, the system. By these reflections we understand that we not only have to reconceive natural history in terms of unfolding symbiosis. That would still be too shallow. We also need to redefine symbiosis itself in a different and more profound way. It is more than a nice form of cooperation between individuals. Symbiosis is a connection that lies deeper than the level on which individuality emerges, an antecedent totality from which individuals emerge and which alone makes cooperation possible. The fact that symbiosis exists points to an underlying unity connecting life and making its forms on a fundamental level identical, parts of the greater whole.

Symbiosis is not only "nice." It means renouncing the illusion of total autonomy. Symbiosis means growth through the other, but for the same reason it signifies continuous death. It engenders the understanding that "I" is much more "thou" than I have ever dreamed it to be, because our existences are mutually interwoven in a necessary and irresolvable manner. Ubiquitous symbiosis makes it possible to understand one another across the different species, across the different animal classes and across the realms of organisms. Understanding is communication of a connective tissue with itself.

This understanding makes it true that "all humans are grass," as Walt Whitman might have expressed it. What in this poetic phrase is a symbol, which the reader intuitively grasps, in the world of poetic biology becomes a deeper truth that connects all life forms. Because my body exchanges its matter with the world, the molecules of the grass also pass through it, like the wind passes through the meadow and makes the stalks wave. I can understand my own body and its stirrings, the deepest meshwork of its roots, only if I experience myself in relation to this other, which is life, in myself. By wading through the fluttering blades in the meadow, I can find their suppleness, their endless ability to answer, and rediscover them also in myself. Both are joined together by one and the same core living gesture, in scale invariance and absoluteness of meaning. What poets link by virtue of linguistic metaphors, therefore, has already emerged as a unity on the level of the body. The task is to track down the universal resemblance and to show that all flesh is of one flesh.

THE "SELFLESS SELF"

If we approach living reality from this perspective, however, one crucial question cannot be ignored any longer: What is my self, if I am not a single player but billions? If "I" is a multitude, is the ego a total illusion, as Buddhists believe? Is salvation from earthly ailments only possible if we overcome self?

If we manage to give suitable, non-trivial answers to these questions, we might be able to alleviate the current confrontation between "ourselves" and "the rest of nature." If we accept the idea of ourselves-as-many, and if we keep in mind the fact that these many selves are inside as well as outside our bodies, then no form of nature-culture or mind-body dualism makes sense anymore. For me this outcome is a promising step forward. But I know that many might take it as a defeat. If a clear self is not to be fenced off, do we not plunge into a bottomless abyss? Is not everything illusion? Maybe the tragedy of our current war against ourselves lies in the fact that there indeed is no solution, but we think that we need to find the key to our dilemma somewhere. We always will be ourselves *and* the others, infinitely pulled back and forth between the two poles.

Francisco Varela, the co-inventor of autopoiesis, calls an organism a "meshwork of selfless selves." For Varela, the organism is without bottom, without a final core. He imagined it rather as a spiral, whose rim is built by the different layers of actors: cells, organs, the body. But in the middle of the vortex, which a living being causes amidst the ever-changing flux and flow of matter, there is nothing. For Varela, therefore, the "actual self" is nothing that an exact biological science could ever find — just as it did not grasp its essence when it declared the "genetic identity" to be such a center. Psychologist Katherine Peil speaks of a fractal self: "Like a set of Russian nesting dolls each looking out from its subjective perspective onto its greater 'not-self' environment (and responding to it locally) which together yields global coherence of the whole. I argue that this is mediated by emotion at each level."[6] The genes active in our bodies and in our lives are not ours, as we have seen, and if we include the metabiome of bacterial and viral genes, then the number reduces to less than one percent. And if we look one level deeper, to the genes responsible for the construction of mitochondria and flagella, we must realize that the heritage we carry around is genes from a whole ecosystem of the past. We consist of histories that are not our history, of bodies that are not our body, but we are only able to exist in this way, as selves that are paradoxically not themselves.

If we follow Varela's thoughts, then we must accept the consequence that our self dissolves into the other and that only the other enables self. In our center we are void like the innermost point of a water vortex. Organisms are so deeply rooted in the other that a living being in its core is made up of that which it is not. If you subtract all the layers of identity an organism actively brings forth in its striving for existence in its desire for plenitude, then the living form itself ceases to exist and only matter remains. In the center of desire waits that which knows no need.

But if the abyss is bottomless, with whose voice are we speaking? Whose pain is it that a living being suffers? And whose exultation to which it tunes in? Is it the desire for life as such that articulates itself? If we keep the universal symbiosis in mind, every individual in some respect is the expression of the whole, which speaks in endless voices. The biological subject, this self-reinforcing ripple on the surface of matter, then, is nothing different than the whole. But it is this whole in the sharpest focus imaginable. Every being is a burning glass, through which all currents, all options of being, can be set ablaze. It is a glowing ember that can kindle sensation, which as a silent yearning may sleep in all matter.

And yet all organic matter is tragically destined to die. The final fusion, the return to the whole, is only possible at the price of death. It *is* death, actually — becoming one with all, matter among matter. The final fusion, therefore, as long as an individual is alive, is only possible in the form of longing, in the form of song. Its echo reverberates from the linden's branches as the voice of the solitary nightingale, which sings as though any night was too short to fully expend its melody. We are more and we are less than we have dared to dream about.

DISEASE: AN ECOLOGICAL DISTURBANCE

In the light of this new perspective we are also able to reevaluate the situations in which the coherence of the individual dwindles. Our usual image of disease depicts pathogenic intruders running up against the bulwark of the body. They menace us as an enemy with his legions threatens a well-armed and fortified country. This view still dominates medical metaphors. According to the perspective thus generated, the threatened body proceeds to an "immune response" in order to annihilate the intruders, in the same vein as today's superpowers reserve a nuclear response for defending themselves. Medical and military metaphors have walked hand in hand for a long time. Both are about attack and defense, about clearly defined friends and foes. If we carefully look at the public education, which the state of Prussia undertook in times of cholera epidemics, we can see that the "inner enemy" of invisible germs was fought with the same means as in the visible world was the archenemy beyond the Rhine.

But such a view, it turns out, misses most of today's problems in medicine. We have seen in Chapter 6 that we cannot explain the role of placebos in healing if we rely on a causal-mechanic pathogen-attack model alone. Now we can add further evidence to this. For example, only very few diseases can be unequivocally assigned to one pathogen. Many diseases of our time, which medicine has not been able to battle yet and that even increase in frequency, are particularly those in which the relationship of our body to itself is disturbed. All autoimmune diseases like asthma, neurodermitis, all allergies, but also multiple sclerosis, are examples of a distorted imagination of other through self. Probably we can include in this view various forms of cancer, and even psychological disturbances, like depression. Additionally, heart and circulatory diseases always have a psychological component. It is striking — confronted with widespread diseases, modern medicine in nearly every case fails to provide a cause or to propose an efficient healing path.

A symbiotic standpoint, however, implies it is searching in the wrong place. From the perspective of the new biology, we can only understand organic disturbances as ecological disturbances. A disease is a partial breakdown of that communication through which the cohabitation of multiple selves is negotiated. If the communication goes awry, the basic process of reestablishing an identity through the transformation of others remains. "Every disease is an unfinished act of creation," the German physician and philosopher Victor von Weizsäcker said.[7] A break in the continuous negotiation tilts the balance between the individuals and the whole. In cancer, a single cell switches into an uncontrolled growth mode at the expense of all others. If we accept symbiosis rather than the struggle of single combatants in a bleak war zone as the principle of biological reality, then disease is not a fight against an enemy but a disturbance of the equilibrium of self-through-others. Pathogens use this disturbance to cancel the former agreement of mutuality and invade the ex-partner. If the balance shifts, disease breaks out. We can observe this by the fact that humans routinely carry highly pathogenic germ lines on their skin, which normally have no destructive effect. They only cause damage after the immune balance has been severely disturbed and imbalances linger over long periods of time.

Many results in immune research meanwhile confirm an ecological view of disease. The healthy body is a well-tuned ecosystem. The number of species included in it, as well as the nature of the relationships among the single species, is in constant change and oscillation. But a fragile balance is always achieved, similar to a landscape in which at some times the rabbits increase more rapidly and at other times the foxes. The immune system does not even clearly delineate which substances belong to the body and which don't. What is outside of the immune self depends on the pragmatics of the moment. What generally is referred to as the immune self — that which the immune system recognizes and defends as self against intrusion from the outside — is not a closed box but rather a way of relating. Even without any pathogen the immune self is constantly bringing itself forth by creating cascades of cells and proteins which bind together, detach and find themselves again. The immune self is as extended in the body as is the neural self, which is represented by the tiny branches of the body nerves, but it has no fixed shape. It is rather a process that is distributed in the liquid expanses of the inner ocean of our bodies.

The immune system is mostly occupied with its own organization, not with defending against intruders. It exists by embracing and caressing self, not by killing it as other. The immune self, we can say, is a mode of self-experience. This holds true for every contact with the outside. Even if an antigen is recognized, the immune cells can only eliminate it if it has become part of self. The pathogen must be ingested by a specific cell and then presented on its surface in order to be recognized as other. For the individual, the other only exists as transformed self. The experience of other is the degree to which the self has been changed by it. Seen from the perspective of the immune self, symbiosis again does not mean nice coexistence but a profound alteration of the self through the necessity of relating.

ALL FLESH IS SUN

What then distinguishes the immune cells, which expand their swimming ecosystem in my lymph fluid and bring forth my identity, from the plankton adrift in the oceans, which brings forth the planetary carbon cycle? What separates the role of the single-celled green algae from the function of the photosynthetic spheres inside the green leaves of the forest, and what distinguishes them from one of my blood cells? What sets the minerals of our bones in the inner ocean of our tissues apart from the chalk mountains built up from seashells which fill the ocean floors? The sea in these musings reveals itself as a kind of oversized volume of a body fluid we share with all. But in the end, the oceanic principle of symbiosis is not reserved to the aquatic spaces of our planet. We understand that the soil, of whose mass more than 90 percent can be organic substance, in truth is a soft skin covering a fertile but vulnerable body.

If we look at the ocean as a whole, its continuous transformations are so much more than the mere variabilities of particular individuals or species in one giant glass jar. At the end of any calculation we have not even touched the living ecosystem itself. It brings forth organisms whose bodies in every detail are the quintessence of their worlds, archetypes of an elusive balance in a cosmos without dimensions. The ecosystem is invisible and boundless, like the viscous threads of the transparent planktonic snails in their sky of blue. The delicate and transparent animals are an image of this interconnectedness, and they are at the same time this relational universe itself. Because all organisms have evolved together in a vast symbiosis of forms, any feature is a hint to another being in the distance, which stays behind the others like a silent shadow.

The floating organisms' forms in this way build up a mirror maze of relations up to the highest level. The blue whale, distant only two steps in the food chain from the energy that nourishes the algae and feeds the krill, is he not best understood as sunlight that has been converted into force and dynamics? Must we not admit scale invariance or fractal self-similarity, as Katherine Peil says, also here, and listen to the lingua franca of our neural system, to this one code that pervades all flesh and that every being understands, which is flesh of this flesh, self of this self? The blue whale establishes the connection. Breathing, striving, yearning like ourselves, he reaches up to the sun. He is the light's tender muscle that unfolds towards the Earth.

CHAPTER 12:

The Silvery Sea

Natural Selection is one of the tools by which Gaia, the self-regulating system, preserves itself as a dynamic but at the same time stable configuration.
— Lynn Margulis and Dorion Sagan

The symbiotic viewpoint opens us a home in the heart of things. The neo-Darwinian opinion, that each individual is a separate atom fighting for its destiny, however, condemns us to loneliness if not self-destructive conflict. If this isolation really was the only merciless truth, as it is still depicted by many scientists, then we could not help but comply. Then it might be best to stoically clench our teeth, keep a stiff upper lip and endure the waves of our vital emotions as the necessary background noise of an egoistical universe. Then, as for a long time it has been the fashion to think, our only hope is that one day our feeling will be unchained from the flesh. We would not be more advanced than the philosopher Plotinus in late antiquity, who clutched to the belief that the "divine spark" of our subjectivity after death would find a home in a bodiless realm of the spirit beyond this bleak valley of tears.

But this valley of tears is, by all means, also a valley of beauty. And the newest findings of biology no longer make it necessary to separate mind from body. The new biology speaks another language. It shows that we are only really stuck if we analyze biological systems apart from their constituents. It shows that a thinker like the French-Algerian philosopher Albert Camus was on the right track when he celebrated the feeling, desiring, mortal and hence finally tragic body.

The current biological viewpoint will change only slowly, however, as long as we accuse the dominant scientific doctrine of myopia from the outside alone. Each major change has to come from research itself. And indeed scientists have begun to elaborate a model of evolution that juxtaposes the known picture of a tree with its divergent branches with a different history of nature. Scientists in that way approach what Gilles Deleuze and Félix Guattari, the famous Parisian thinkers of the 1980s, would have called a "rhizome." This new history of unfolding freedom corresponds to a subterranean meshwork of roots and fungal mycelia with its redundancies, confusions and shortcuts — and with its anastomoses, those sprouts that miraculously find one another and instead of remaining alien to one another, keep on growing together.

Scientists have started to trace these convergent branches throughout natural history. They call the hypothesis "reticulate evolution" — the idea that evolution is, at least partially, best understood not as a linear history but as an unfolding network. Proponents of reticulate evolution hold that species do not diverge for once and all time by outcompeting each other in the evolutionary arms race but often join to form new unions. They claim that this process of joining lifelines is a powerful means of making new species come into being. Reticulate evolution, they think, allows the quick creation of fully functional new life forms in addition to existing ones, whereas in classical competitive evolution, one species gradually transforms itself over time without bifurcating its home branch on the tree.

In the view of reticulate evolution, the borders between the species are much more blurred than neo-darwinism has supposed. This permeability is made possible by a phenomenon that until recently was regarded as impossible: unexpected cross-fertilizations between two species, in which sperm and eggs from remote groups merge. This is a revolutionary stance for a picture of nature that has so exclusively adhered to the image of the tree with its linear growth and neat separations of the distinct branches and twigs. Now a host of evidence makes it probable that there have been many fusion events in life history. Though the extent to which this might have happened is still under hot debate in the scientific community, there is a lot of evidence that species have continuously crossed during their development, and by this have exchanged many of their features, behaviors and shapes in order to yield totally new heritage lines. Not only single-celled organisms, but also metazoan lineages inside the groups of such highly developed taxons as insects and sea urchins, so some scientists claim, formed new junctions by this process, with novel genealogies emerging from their lifelines.[1]

I was allowed to see indications of reticulate evolution with my own eyes. As you might guess, this also occurred on the ocean. It happened during a cruise on the research vessel *Heincke*, which in many respects became a journey into the womb of other life and its unfathomed depths. The cruise took place in June. We did not go further than the German Bight, a coastal sea with ever-choppy gray waters, one that extends a hundred odd miles to the west of the German North Sea coast. During the whole two weeks of the journey, a chilly northeastern wind swept the sky clean and made our vessel roll in a long smooth swell. We were hunting less for animals than for chemical substances. The cruise was meant to assess the distribution of agricultural nutrient runoff in coastal waters. We documented the creeping transformation of the North Sea into a well fertilized fish pond.

The vessel's movements were accompanied by the hum emerging from half-filled refrigerators and the clanking tubes and metal clamps of the chemical equipment. I will long remember the smell of saltwater and the scratched lino floor of the ship lab. I was often at work in the night. I shared the night shifts with a good-humored middle-aged laboratory assistant. Laconically, she always murmured the same word when the vessel was hit by a larger than usual wave amidships and we slid some yards over the floor on our chairs. "Nice hit," she repeated with a slightly tortured voice and then pulled her chair back to her desk. In the first nights we both were basically alone, because many of the science crew had had to give up working due to seasickness. After a day and a night of travel with a nice strong backward gale the ship had arrived in the sample area and veered 90°, so that the seas suddenly came from athwart and the ship immediately started to roll in a quite unpleasant way. Soon everybody had vanished, even the sailors.

After sailing from Cuxhaven, we made a brief stop at the island of Helgoland. This small stretch of red rock is the only German island in the high seas. I seized the opportunity to borrow a long plankton net from the Biologische Anstalt, the marine ecology institution on the isolated patch of land among the greenish waters. As soon as I had the net, I started to drag it beside the vessel for some minutes after I had finished my work at the sample station and some spare time was left. I stood on the wooden deck planks amidships and watched the metal ring, which kept the mouth of the net open, dance in the swell. The whole thing seemed like a soaked long gown that was spread and rocked by the waves.

We wore heavy boots and had pulled the trouser legs of our bib and braces over their rims in order to keep dry whenever a huge wave washed over the deck. When the vessel was ready to sail after all the station work was done, Jürgen, the head sailor, hauled in the net and let it dangle above the deck so that I could rinse its meshes using the seawater hose in order to concentrate my catch in the small alloy bucket at the lower end of the mesh. I had to start processing the chemical water samples first. After I finished this, cleaned all the glass gear and was ready for the next station, I had some time for my own catch. I poured a portion of my prey into a glass dish and pushed it under a powerful stereo microscope in the ship lab.

I will always remember the moments that followed.

It was early summer, so larvae predominated, undulating through the water. They showed me that we were still in a coastal sea where the ocean floor is not that far away from the surface. In its mud dwell a diversity of life forms, a menagerie that in spring procreates in a fertile explosion. It happens all at once. The sperm and eggs of the majority of animals are released during a very short time span into the open water, where they eventually meet and merge into a larvae. Under my lens I discovered revolving spheres, which were covered with delicate little hairs. They populated the water like germs with the potential for universal development. I knew that some of them would transform into marine worms; others, however, which now looked indistinguishable from the former, would become snails. In search for more clarity, I leafed through my *Newell*, the most useful of my beloved plankton field guides. I still have all these books on my shelves at home, and when I open them, the faint smell of saltwater and formaline reminds me of these nightly hours on the rocking vessel.

While I compared the drawings of the rotating spheres, the *Trochophora* larvae, I had the impression that this single form offered a kind of universal first body for a host of animal species in different groups, which are not directly related, like worms and snails for example. I discovered different kinds of such generic larvae in the swell beneath the lens. These animals had a highly transparent body, with shapes I could barely discern from the surrounding water. These larval forms carry the romantic sounding moniker *Pluteus*. The translucent Plutei faintly reminded me of bloated gummy bears that had stayed overnight in a water jar. This type of precursor form can give rise to the adults belonging to several sea urchin and starfish species. So some virtually indistinguishable larvae yield adults of different animal classes. On the other hand, and this felt a bit strange to discover, some closely related species which as adults looked rather similar, have totally different larvae. Still other species entirely renounce the chaos of larval precursors and hatch directly from the egg, for example, as a minuscule starfish.

Right in front of my eyes a tiny extract of life moved back and forth in the rhythm of a huge breath, one that lets the surface of the oceans rise and sink again. I then asked myself, which of the many different appearances during an organism's life cycle was more real? The larva? Or the adult animal? After all, in a lot of species the juvenile forms last many times longer than the adult ones, who often are granted only a brief mating ecstasy and then quickly fade away. An extreme example of this imbalance is the life cycle of the North American 17-year cicada. The larvae dig their way in the soil for year after year, until in one summer crowds of young adults in mating fever show up all together. Chafer grubs, the larvae of the common European May bug, which swarms in late spring, spend years in the soil eating roots before the adult is granted some brief nights of warmth in the tree crowns. The mayfly crowns the months of preparation during which the larva has been lurking in the water, its body pressed to stones, with a jerking mating dance for a few hours around sunset. These one-day-stand lovers are not even able to ingest food anymore. But a long larval life is not the only possible confusion in the logic of reproduction.

There are also species where the adults are anticipated by several different larval forms that can look entirely distinct. This is particularly marked in the shrimp species of the genus *Sergestes*, which with their shiny red color float through the darkness of the mid-ocean waters. Here the developmental whim is not satisfied with one single larval precursor form, nor with two or three. The development of an adult *Sergestes* shrimp must pass through the bottlenecks of five moltings, wherein larval forms are so different from each other that they resemble five distinct species that transform into one another. During all these metamorphoses, accidents happen regularly and subsequent mass deaths occur, as if the separate clothes did not really fit over each other and suffocated some of the individuals desperately trying to get rid of them during molting.

Are adult forms usually regarded as the real representatives of a species, because through them the kind is continued? The Greek philosopher Aristotle would have probably said that only in that developmental state that is able to reproduce is the purpose of the species fulfilled. But the adult is as enclosed in the larvae as reality is enclosed in a dream. For larvae that shed one sheath after the other, therefore, the Delphic oracle's exhortation to "become who you are" seems all the more valid. On the other hand, the true being, which our goal-directed minds so easily confuse with a mechanical purpose, might not lie in the finished product of all these juvenile transformations but rather in the metamorphosis itself. Here we come back to the dialectics of self and other, where the other hides in the self and vice versa, and only both together yield the whole form.

Mainstream evolutionary thinking, however, believes it has found a good functional explanation for the dual lives of many invertebrate animals. A developmental history in which one species splits up into two ways of life, many researchers argue, simply pays off. The young do not devour their parents' food because the adults thrive on a different diet than their offspring. Caterpillars chew leaves; moths suck nectar. Never put all your eggs in one basket. By investing in several strategies, a species, so the theory goes, minimizes risk and has well spread its portfolio.

Nonetheless many questions remain open. What use is it for the shrimp *Sergestes* to undergo huge losses during five metamorphoses? In *Sergestes*, the ways of life of the distinct forms do not differ (they all feed on planktonic algae), so there is no advantage in this respect. How then can we justify the extreme expenditure necessary in order to break down a functioning body and to build it up again from scratch? With what value can we account for the half hour of total helplessness a freshly hatched butterfly encounters before its wings are dried and the body hardened? Why do the *Trochophora*-type larvae of worms and snails resemble one another so closely, although the larval stages of these species thus compete with one another, losing the advantage of reduced competition to their parents?

ALL BUTTERFLIES ARE TWO SPECIES

The British morphologist Don I. Williamson, who works at the marine biology laboratory in Plymouth, is one of the most zealous proponents of reticulate evolution. He is not satisfied with the linear model that needs to invent a payoff function for even the weirdest body feature simply because usefulness is the only explanation possible in mainstream theory. For years, Williamson has been working to posit a theory that can explain the observed phenomena and does not need to resort to infallible dogma. The idea Williamson came up with is so fantastic that his colleagues have been laughing at him for nearly three decades. The Briton's idea works, but in order to fit it into the building of biological thought, we need to give up a central pillar of evolutionary theory.

Williamson thinks that all species that contain a larval stadium notably different from the adults, in reality are not single species but at least two. Juvenile forms and adults, Williamson holds, originally stem from different branches of the evolutionary tree. Their chromosomes became shuffled because sperm of one species managed to fertilize the eggs of another. In a host of different larvae like *Sergestes*, twisted developmental pathways are simply the result of these multiple overlaps. They have come into being and just work without being particularly efficient. In the sea, scientists have observed, the meeting of sperm and eggs of different species is not a rare accident but rather the inevitable daily reality. In springtime, countless beings open their reproductive glands into the water, which then becomes a thick soup of first germ cells and then embryos. Every egg is open to be fertilized by whatever germ is able to pierce its wall. Unforeseeable chimera are, therefore, not excluded. Once such a fusion happened, Williamson thinks, it might have become retained, not because the emerging life form was more efficient, but simply because it was not impossible. The advantage of existing in two niches — feeding on plankton as larvae and on bigger animals as adult — may have given some additional momentum, but was not responsible for the existence of the novel chimera. According to Williamson's model, the new being was created by chance. It was made possible by the astonishing flexibility of an organism's developmental modules.[2]

Only today, in the face of the radical change that our view of embryonic development is undergoing right now, is Williamson's explanation of the animal's whole-genome metamorphosis beginning to seem plausible. If we add to this the evidence that we are living in a totally symbiotic biosphere, and if we supplement this by the latest insights into how far organisms are free to interpret DNA signals during their development, then reticulate evolution follows nearly inevitably. But for mainstream thinking it has remained a very controversial idea. In the plant realm, however, botanists have observed a host of hybridization events. Here, an evolutionary view that includes mergers of lines has become widely accepted. Many plant species have arisen when whole genomes of different original species merged, as for instance in rye.[3] Hybridization of genomes is even a common tactic of plant breeders to yield promising new strains. But as widespread as this mechanism is in botany, many think it to be improbable in animals.

Williamson, however, has managed to show in his lab that fertilization across species barriers is possible and that the emerging embryos are able to survive. He united the eggs of a sea urchin with the sperm of an ascidian, or sea squirt, a distant relative of the phylum and genus *Homo* belongs to. The resulting chimeras did not die; to the contrary, the eggs developed, and some days later hatched tiny sea-urchin-larvae (of the *Pluteus* type I had observed under my microscope in the rolling North Sea). So the development started with the larval form typical for many sea urchins. But then, Williamson could not believe his eyes when the Plutei after a while converted into minuscule ascidians (sea squirts) and settled to the bottom of the aquarium. In principle, this experiment shows that nothing impedes cross-fertilization not only across species barriers but also across whole biological phyla. "Probably at the beginning there were only species without larvae," Williamson believes. "These latter were all acquired later."[4] He thinks that those starfish species of today, which directly evolve from eggs into the adult shape, are the original forms whose close relatives have acquired a larval stage through horizontal gene transfer. Some of these cousins then split up into different lineages through the incorporation of other bodies. These often arose by crossing with one particular life form that closely resembled their actual larvae. Therefore, the same few planktonic precursors may have been integrated into different lineages several times. This could elegantly account for the fact that many marine worms and snails seem to have an identical larval form, the *Trochophora*. "The same larval type has inserted itself into different phyla," the symbiosis researcher Lynn Margulis also believed.[5]

The progenitors of such natural experiments presumably still live out there. Much speaks in favor of the hypothesis that the *Trochophora* is related to modern rotifers — minuscule rotund animalcules that inhabit freshwater and saltwater and are quite ubiquitous. When I was a kid and loved to play with an old microscope, I used to extract rotifers from the water of flower vases my mother had not bothered to change. I could look for hours at their peristaltic dislocations in the small Eden under the lens.

There is more evidence for these kind of marriage in the animal realm. Take coelenterates, which include medusae, polyps and corals. Contrary to what marine biologists previously believed, many jellyfish do not directly hatch from eggs but emerge from specialized polyps. These tiny, fleshy stalks cut off minuscule medusae from a special organ on their upper side. The baby jellyfish detach into the ocean water like small soap bubbles from a straw, and then start the life cycle anew. Medusa and polyp can, therefore, be interpreted as the adult form and the larvae. The Russian biologist Ilya Ilyich Mechnikov, who set up a research lab in Messina, Sicily in the 1890s, observed a strange intermediate form between polyps and medusae, in which bodily features of both seemed to be stuffed together in one body. Each form moved its limbs without any coordination with the other.

If Williamson is right, the history of evolution needs to be rewritten. The tree of life not only has diverging branches but also many which grow together to form new lines, and through this process fuse groups that are on rather distant branches of the evolutionary tree (which through this process becomes, instead, a dense shrub). Provided the hypothesis of reticulate evolution holds true, we will have to accept many biological strangers as close cousins. Many scientist think it possible that such a cross-fertilization was the reason for the explosion in fossil diversity archaeologists observe in Canada's Burgess Shale formation. There, it seems as if the whole diversity of today's phyla had been created out of nothing, and wild new body plans abruptly appeared in the fossil record: marine animals with spines on their back like the elusive genus *Hallucigenia*, swimming creatures loosely resembling crustaceans and many forms of worms. Lynn Margulis believed that "the huge developmental shifts in invertebrate animals have come about by the transmission of wholly new genomes."[6]

All larvae of insects that undergo a complete metamorphosis (as do butterflies, in contrast to cockroaches), Williamson and others think, have emerged after a cross-fertilization with velvet worms, or Onychophores. These tropical animals in many respects resemble caterpillars: they are worm-like, have a soft skin, plump legs and glands that produce a sticky liquid. These glands in particular are a feature, Williamson points out, that adult insects do not own; only their larvae have them. The silkworm, for instance, uses the glands to construct cocoons. Adults of the tropical weaver ants use their young as living yarn reels in order to sow their nests from foliage. The ants clutch the larvae with their mandibles and stitch the leaf parts together.

For Williamson the presence of silk glands in caterpillars is evidence that the velvet worms are interspersed into the insect lineage. Seen from this vantage point, we can reevaluate some efficiency calculations. For instance, the fact that weaver ants cannot rely on proper silk glands, but instead need to use their juveniles as living stitching needles, does not seem very parsimonious but rather seems an inconvenient but somehow functional mixture of life's coincidences. There seem to be at least two very dissimilar larvae types in the insect kingdom, the caterpillar and the maggot type. In the huge group of Hymenopterans, to which ants, bees and wasps belong, the maggot type is widely distributed. Only the sawflies, also hymenopterans, differ. Their larvae look strikingly like caterpillars. This discontinuity is an argument in favor of the insertion of different precursor lines into the insect realm.

Williamson's idea could be an elegant solution to the problem that often the empirical proofs speaking for a parental relationship among species are not without contradictions. Often, a lot of features for a species suggest membership in a certain group, but some others speak against it. This leads to a continuous regrouping of family trees

— the play field of taxonomists, those biologists who are particularly obsessed by the supposed order of nature. With Williamson's theory in hand, one can imagine that all such contradictions are real because they mirror the varied history of coevolution and its diverse cross-breed experiments of merging and fusing and separating again.

Genetic tests have started to show that indeed the old textbook classification of many taxonomic groups does not hold up to genetic studies. With the aid of genetic marking methods, the DNA of very distant groups can be compared and common sequences can be found that are evidence of some parental relation. But by no means do all genetic studies show consistent results. Indeed, some genetic analyses yield support for reticulate evolution, although others show the contrary.[7] Some regrouping of the invertebrate evolutionary tree done through molecular analysis shows that the caterpillar-like velvet worms, which so far had been put on a quite remote dead-end branch of the tree, are rather close to the insects. And the annelid worms, containing the humble earthworm, which for generations of biologists was thought to be somehow a precursor of all insects, have been placed farther away. On the other hand, in some of these analyses the rotifers from my mother's stale flower water, have been placed close to the phyla that pass through the stage of a *Trochophora* larva during development. According to these genetic studies, the inconspicuous animalcules from the vase water have come much closer to their more dignified relatives, from which biologists had distanced them for the reason of exterior appearance alone.

All this means that biology students who are sitting in the lecture theatres right now are learning an outdated systematics. Already huge chunks of the old knowledge that I had to memorize during my undergraduate biology studies have been dismissed — like the coelom hypothesis, which grouped animals according to the way their body cavities were formed during early embryogenesis. The great thing in science is the fact that what objectively seems to be true and irrefutable today, inevitably will be tomorrow's blatant error. The relativity of knowledge can cause shivers. Those who propagate ideas that in the future might become the new mainstream, however, are mostly beaten up by the establishment. But this is the fate of all revolutionaries. The innovative potential lying dormant in the idea of reticulate evolution might only be comparable to Carl Linnaeus's *Systema Naturae*, which in the 18th century became the basis for all modern attempts to order the multitude of life forms. Linnaeus was also a revolutionary. He helped put biologists on the track of evolution, as his system introduced the idea that different species are naturally related to one another and have not each been created independently by a watchmaker god.

So why then the maggot and the hornet? The caterpillar and the butterfly? The *Pluteus* and the variety of starfish and sea urchins? The ubiquitous *Trochophora*? In the light of reticulate evolution a complicated developmental path is never chosen for reasons of a perfect adaptation. It rather is a useful accident. It is a chance event that follows from nature's obsession to produce novelty at any price and then to play with it. This means that metamorphosis is a particularly strong case of the reuse of already functional principles. "Nature is working as a tinkerer with available materials," as François Jacob, one of the discoverers of bacterial gene regulation in the 1960s, has stated, "and not like an engineer who is always inventing everything anew."[8]

NEW CHROMOSOMES BY COPY AND PASTE

Geneticists are debating a second mechanism, which might be able to abruptly create biological novelty. Karyotypic fissioning is an event occurring inside the genome, by which genes suddenly change their order because single chromosomes, the elongated agglomerations of DNA strands in the cell nucleus, break into several parts. Such a separation can happen, for instance, when the center spot of a chromosome, the centromer that keeps the chromosome's arms together, suddenly doubles, yielding two smaller chromosomes instead of one larger. As all genes remain untouched, at first this process has no consequences. No gene is lost and no gene is doubled, even though they are physically relocated. The change, however, becomes significant during procreation, when, for instance, a sperm cell with the newly formed smaller and more numerous chromosomes fertilizes an egg cell with the normal number of chromosomes, changing the number of chromosomes in the embryo. By this process, certain genes enhance their influence, while others vanish completely. This can have far reaching consequences for the developing embryo. If it survives the new configuration of its own DNA, anything from a slight shift in traits to a huge restructuring of the body can happen. In any case, the embryo is extremely sensitive to any changes in the developmental sequence; slight differences in timing during development have been shown to result in an altered phenotype. Therefore, karyotypic fissioning is a highly potent instrument of major change in the shapes of organisms.[9]

Using the hypothesis of karyotypic fissioning, evolutionary biologists could elegantly explain many leaps in natural history and particularly those of vertebrates. Some researchers argue, for instance, that this mechanism is at the root of primate diversity on the island of Madagascar. On that isolated landmass off the coast of Southeast Africa, lemurs have developed, conspicuous tree dwelling primates with huge dark eyes and skinny fingers. But lemurids also exist only here, a closely related group that is nevertheless distinguished from the lemurs through some distinct variants. Both groups occupy overlapping niches, and each of them has a specific but different number of chromosomes. It is possible, biologists think, that by a single fissioning event, the original lemur precursor species split into two lineages with different chromosome numbers. By this process, we could perhaps also elucidate how contemporary man has emerged in leaps from the pool of his ancestors. Great apes have 48 chromosomes, humans have only 46.

THERE NEVER WAS A MISSING LINK
All these theories, including Williamson's wild idea, are not fully accepted, at least not yet. But they are at least being discussed and experimentally tested. This is a big step forward, compared to the

ultraorthodoxy that only 15 years ago silenced nearly any alternative approaches to organisms. In the meantime, embryologists have observed how flexible a developing being is, how far from being a mere slave to genetic orders, which were once thought to contain all conceivable necessary information. Scientists have understood, as we have seen in Chapter 3, that an embryo does not execute genetic orders during development, but is rather related to its genes by weak linkage and not by causal mechanics. The embryo is free to interpret within a certain range. The developing body deciphers all incoming impulses in relation to its need to keep a closed self-producing unity. Incomplete areas are thus automatically elaborated in a meaningful sense, without needing to be coded in the genes, as for instance newly-built bone tissue self-organizes in relation to appropriate nerves, muscles and blood vessels. By this flexibility whole genomes can be integrated as well. We need not to imagine chaos and accident, as the body is made to work as a meaningful whole out of unrelated impulses. As we have seen above, the genes do not provide a strict set of orders, but rather an inspiration for new imaginative possibilities. Not every gene must be tuned in to every other. The body, intending to preserve its unity, integrates them into a functioning whole without needing any concrete set of instructions in order to bring forth a living being, which experiences its existence in meaningful ways.

The Israeli biologist Eva Jablonka believes that, "Evolution's menagerie answers much more immediately to the environment's influences than the doctrine of chance mutation wants to admit until today."[10] Organisms come to meet the forces of selection by active imagination. Their body's self-construction is able to translate weak genetic impulses into new wholes by virtue of its self-forming powers. They do not have to wait until a whole genetic regulatory cascade has formed. Some genetic change is enough in order to inspire the automatic developmental processes able to create meaningful forms and to integrate the new note into the score. Also in this respect organisms are not a perfect molding of the features of their environment, which controls its inhabitants by the constraint of survival

311

of the best adapted. To the contrary, autonomy plays a much bigger role than previously supposed. Novelty arises from the negotiations between living agents and the necessities of a habitat. It is a mutual give and take, a reciprocal creation. Katherine Peil observes: "This is the ongoing feedback cycle between self and its environment, where each continuously defines the other, and emotion signals self-relevant changes to which the organism must adapt."[11]

Orthodox Darwinians had assumed for a long time that novel species arise through a long sequence of gradual steps. New features start as barely perceivable, and become reinforced if they are useful, or so the theory goes. But very few intermediate forms have been observed like Archaeopteryx, the reptile with bird feathers. And it is totally unclear what the path which led to Archaeopteryx might have looked like, as well as the one leading from there to modern birds. In any case, it is improbable that subtle changes could have conveyed much advantage to the individuals that bore them. To the contrary, it is difficult to assume how efficient and how successful these living alpha versions of novel evolutionary ideas could have been. Biologists have come to call these mostly hypothetical forms "hopeful monsters." In the light of what we have observed in the previous lines, it does not seem at all improbable that all these intermediate types have come into being through major chromosomal rearrangements that allowed entirely new forms to appear in one fell swoop.

As for my story, after a couple of days we had to suspend our sampling work. On our last night, when the waves rose gradually higher, Jürgen, the head sailor, had muttered between his thin lips: "Soon the research will be over here." He was right. The wind first shifted to the south and died away during the night, while a steady high swell was tossing us around. Then the wind turned again to the west and quickly grew to gale force. Its strength drove the digits of our anemometer, which was bolted under the ceiling of the ship lab, to new records. The *Heincke's* bow swung up and with a thundering roar fell back into mountains of a steaming blue, which then boiled up to the windows of the bridge. The sky was washed clear, polished like a wet pebble in the surf. We surfed down blue ridges with our 1,000-ton vessel as if the *Heincke* was a giant sled. But soon, doomed by the weather to inactivity in the open ocean, we retired into the Elbe estuary in order to finish work on some of the remaining stations and to wait.

Yet that night before the storm, the sea had bestowed me with one of those amazing experiences that I will never forget. The moon had come out and poured so much light over the ocean that we could have switched off the deck lamps. It was a *mar de la plata*, an ocean made of wrought silver, like the bodies of the fish in that swirling underwater cloud off La Gomera. The world had turned into a glistening relief of itself. As I was waiting for the return of the CTD probe in order to take my water samples, I could see that a movement stirred the surface; it looked like the bursting of innumerable tiny bubbles. The sea was seething like my small rural pond, which had first taught me to see. I looked closer. First I thought that the creatures moving through the water with wriggling movements were fish. Maybe eels. But then I realized that the animals dancing at the surface were worms. They were shining red and big, some of them maybe double the length of my index finger. Hundreds, thousands, millions of worms swirled through the body of the high sea. They had left their burrows in the sandy ocean floor and met in the upper water layers for their mating dance. The moon had lured them all up at the same time so that, unfailingly, they would meet. They swarmed around one another, danced and wriggled, spiraled and swayed, until they released the eggs and sperm from their bursting bodies and then sank back powerlessly into death. It happened secretly in the dark sea, an ecstasy of colors, meant for nobody's eyes. The fact that I had seen it was pure chance. I asked the first mate to switch on the huge searchlight on the bridge and direct its beam into the water. The powerful light pierced deeply below the turquoise green surface and gave an untarnished transparency to its glassy volume, one that was only interrupted by the circle dance of the worms. They populated the ocean as blood cells inhabit a living body.

Part Six: Healing

In this last section I will explore the ethical consequences of our deep entanglement with the rest of creation. I will investigate the viewpoint a poetic biology implies for an ecological ethics. Until now all attempts to find moral rules for the protection of other species have remained rather theoretical. Most attempts have had a common vantage point: man is opposed to nature or at least stands outside of it. Therefore, most ecological morals try to entice us to become a good steward — to control the planet from a supposed outside, as we are already doing, but in a more benign way. A poetic ecology provides us with a radically different picture. If the borders dissolve, if we are nature and nature is us, then ethics becomes a very different endeavor. Then it is no longer about preserving useful goods or protecting others. Acting morally towards others is not conceivable without acting morally towards oneself. The ethics I will explore follows the idea that life bears values according to its vital needs. All living beings can understand them. Preserving nature, therefore, is about preserving our own identity. At the same time it is about protecting what is totally alien to us. Both can only be achieved in a joint manner. Preserving nature means to remain healthy in a healthy environment. "Healthy," however, does not translate into whole and undamaged but rather into the freedom required to carry out necessary embodied imagination. This poetic imagination is a requirement of a healthy system. It is, if present, something that has a healing effect because it instills aliveness. Aliveness means to be able to creatively participate in the ongoing imaginative processes in an ecosystem. Historically, the rules for such a mutual participation and interpenetration have been developed through the commons, where humans co-creatively interact with nature in order to allow all participants to flourish. An environmental ethics, therefore, must be a set of commoning rules for ecosystems.

CHAPTER 13:

Ethics:
The Values of the Flesh

*With all his knowledge about matter, he [modern man]
is ignorant with regard to the most important and
fundamental questions of human existence: what man
is, how he ought to live, how the tremendous energies
within man can be released and used productively.*

— Erich Fromm

8:42 a.m. With the scalpel we draw a rose-colored furrow into the scalp of the rabbit. The smooth tissue unclasps with a mild scrunch and sets the white skull bone free. "Drill," the professor orders. Ines, his assistant, hands him the device. Before the professor places the spiraling metal tip in position, I pull aside the rims of the scalp using a small, alloy rake. It is hot down here in the basement. The shelves covering the walls are cluttered with folders and old, disused electrical measuring apparatuses. Ines, wearing a t-shirt, is sweating. Drops cover her forehead. She does not say a word. The drill makes a sound as though it is burrowing through a thin plastic sheet.

The surgery protocol of March 6, 1992, records that we had anesthetized the rabbit half an hour before with Ketanest and then inserted the respiration tube into the animal's windpipe. This was the most delicate step in the whole procedure. We had to bend back the flaccid neck of the helpless animal and then thread the thin, rubber tube past the four long incisors so the end of the tube could slip into the trachea. We were proud to be the only neurophysiological research group in Germany capable of intubating a rabbit as expertly as the well-trained staff of an emergency room.

But unlike surgeons in a hospital, we were less equipped with shiny instruments to monitor our work. We particularly lacked a device to measure blood pressure and heart rate. The last time, the rabbit died shortly before surgery. Today, the professor has ordered not one, but two individuals from the animal facilities of the faculty. You never know. The other rabbit is still nibbling dried grass pellets in a plastic box in front of the lab door. But so far, everything has gone well. After the successful intubation, we injected curare to slacken the animal's muscles so we could work in a more relaxed way. This was why we needed to attach it to a respiration machine in the first place. Otherwise, it would suffocate. Curare blocks the chest muscles (the Amazonian Indians use this substance for the arrows of their blowguns). If the rabbit moves even a millimeter when we open the brain, our electrode would miss the few, single brain nerves the professor wants to isolate. We have clamped the animal's head into the stereometric apparatus — a sort of alloy scaffold with several metric scales to measure the position of the electrode — bearing a smooth screw thread to push the tip into the soft brain tissue. The respiration machine fills the room with its rhythmic splutter.

8:58 a.m. The skull bone is severed. The professor took a painstakingly long time to abrade the last layer, thin as paper, without bruising the brain. Now the fine basketwork of the outer meninges, the stiff, moist skin surrounding the brain, is visible. On its surface, I can see delicate streaks of dark threads and irregular, black dots. The smooth top glistens with moisture. The professor wipes sweat off his forehead and gazes searchingly into our faces. Why are we so tense? Until now, everything has been as easy as assembling a Lego kit. The breathing machine hisses in utter monotony. The rabbit, covered with sterile green cloth, with the tube in its throat, is stretched on its side, its skull looking less like a wound and more like an absurd construction site.

The professor begins to readjust the electrode, which protrudes from the stereometric apparatus like a tiny toy crane of shiny metal. Ines switches on the oscilloscope. Blue flickers of light fill the room. If the fine electrode wires hit functioning neurons when they are pushed into the tissue, we will spot a regular pattern on the screens. If we are lucky, the monitors will show the typical discharge images of nerve cells. The closer the electrode is, the more coherent firing we will see. The dose of curare in the rabbit's blood will last for another 47 minutes.

CRUSHING THE IRRESISTIBLE MAGIC OF LIFE

I had assembled the fragile probe in the months before. I was in the fourth semester of my undergraduate studies in biology and worked as teaching assistant for some extra money. The surgery was meant to test my construction. With no multiterminal electrodes for rabbit brains available on the market, we had to create our own from parts we could find in electronics stores. Finding the components was not difficult, but assembling them was. I was sitting in the attic of the neurophysiological institute, formerly a private mansion, trying to solder a construction that functioned reasonably. The warm air of the Freiburg summer evening streamed through the windows while light seeped out of the room. I could smell the summer warmth: the fir trees in the park and the promises of the night — I knew some friends were having drinks in a beer garden. I floated on a wave of scents reaching back into the 1850s and the happy times of the German university and its student life.

The wires were only nanometers thick, with a golden color, reminding me of the very fine hair of a child. I had to strip off their insulation and thread them into a plastic plug where I could solder them to tiny copper tubes. The wires were supposed to stick out of the plug like hairs of a tiny brush. This plug would be cemented to the skull bone once the electrodes reached an interesting signal in the soft matter of the brain; that was the aim of the surgery. On the other side of the plug, designed to protrude from the skull, a common socket could be fitted that connected the electrode to the control monitors. It would be very handy to have a lab animal that could be easily plugged in. In fact, we were constructing a cyborg, a new individual that mixed animal and technical attributes. This rabbit cyborg was to allow us deeper insights into how the visual cortex of the brain worked. Perhaps the professor was already dreaming of

THE BIOLOGY OF WONDER

some major paper. But he had not taken into account the difficulty in constructing the necessary electrode and how difficult it would be to sink this into the warm, soft, living brain.

My work with the probe advanced slowly. The wires broke all the time. When I managed to insert a few of them into the different copper tubes, it often happened that I applied too much heat while trying to solder the next wire, and everything came off. I had to start over from square one. It was a real Sisyphean task. I had the feeling of sitting there for months, undoing each day's work in the evening, like Penelope with the weaving on her loom as she waited for her husband Ulysses. I was sitting in my attic, beginning every evening from scratch, and breathing the sweet air that was heavy with summer and restless with youth. For a while, the professor did not ask about my progress. I did not tell him until he needed the plug, and then he became very impatient. For half a week, I toiled day and night and still only had a partially working prototype. The probe he ended up inserting into the rabbit's brain by means of the stereometric device was not even half of what he had hoped to use. Instead of the 36 working electrodes he wanted, it had only 13.

Still, I did not consider quitting my job. Not that summer. I wove my absurd creation of copper hairs and hoped it would lead me to more knowledge about the living. I wanted to *know*, and I believed in the professor, with his unkempt hair, reading glasses down on his nose tip and sparkling wit that emerged in his remarks from time to time. I would know, if only I followed him. He was an outsider after a great idea, or so I thought. I considered him to be rather benevolent. That summer, I did not make any connection between what I did and the wailing dogs in the basement of the faculty of medicine, which I heard every time I walked to the students' restaurants with my buddies. Their barking sounded like the fruitless screams of medieval prisoners in dungeons deep under a tower.

I did not make any connection with the sparkling, green tiger beetles in my secret path in the Kaiserstuhl hills, southwest of Freiburg. They lived on a trail between the high walls of soft white loess with wild grapevines hanging down from the sides, where I could not help but walk barefoot to feel the utter softness and warmth of the sand between my toes. The tiger beetles also seemed to love the soft, warm sand. This path was my secret rescue place when I was feeling down. When I went there one late summer afternoon, they had started to tar it all over. The beetles were not to be found. It smelled of bitumen. Machines were parked among the vines. I did not go back for years.

But I did not make the connection then. I wanted to know what life was and what sat in the center of its irresistible magic that lent the world its melodiousness. I believed at the time that I was on the right track in my medieval garret among the summer scent of the firs.

9:03 a.m. The professor has adjusted a lens above the hole in the skull of the rabbit. Ines turns the ratchet knob and starts to lower the spindle carrying the electrode. She has connected the plastic plug with my twisted wire brush to a cable. The monitors quiver. The professor, looking through the eyepiece of the magnifying glass, waves his hand like somebody directing a truck to the loading ramp. Ines turns the ratchet. The respiration machine trots along in its slow staccato. "Stop!" the professor screams. "Stop! Hang on! Back, turn back!" Ines stops and turns the ratchet with a jerk. "Back, I said!" Ines turns more quickly. She has lost the direction. Sweat runs down her neck. The professor jumps up and seizes her hand. "Have a look at that," he says with a flat voice. We look through the lens, first Ines then I. The golden wires have bent to all sides on the hard meninges. Some even seem loose. And we are still far from being inside the brain. The professor leans on the wall and watches Ines angrily. "Go do it yourself if you don't like how I am doing it," she hisses. The air in the basement is stifling. We are as nervous as a military command post under fire. I start to carefully wind back the electrode. "Don't touch it!" the professor yells at me. Ines proposes to incise the meninges with a scalpel. The professor disagrees. He does not want to risk any additional lesions. The bent wires tremble above the wound in the animal's head. Its eyes are half open, as if it perceives everything happening but cannot move a fiber of muscle.

321

The respirator wheezes its mechanical beat. The curare will paralyze the rabbit for another 42 minutes.

VIVISECTION HAPPENS EVERY MINUTE

Nobody prohibited what we did; on the contrary, the central animal facility of the university had thousands of beasts ready to provide an uninterrupted flow of living flesh for thousands of scientific experiments. Today, specialized companies breed living beings of any kind, providing geneticists with any conceivable mutant, virtually ready-made. The Jackson laboratory in Maine, for instance, ships its Jax® Mice worldwide. The rodents are bred in a host of strains with any desired genetic defect.

Our final goal, however, was not to remain stuck with rodents. We were aiming higher. The rabbits' skulls were only a test for more ambitious ventures. Well-hidden in the park behind the institute building was a modern compound with tiled floors and walls, containing several huge cages with clean chrome bars. Here, we kept rhesus and southern pigtail macaques. These were our real guinea pigs. They were to be spared until our methodology was perfected. The professor wanted to avoid an expensive monkey dying through imperfect techniques, which meant we had to dissect rabbits for an indefinite time.

In the meantime we trained the macaques. To do this, it was necessary to carry the animals from their stable to the lab. We had to push them into a transport box, carry them through the premises to the basement where the lab was situated and then fix the animals in their monkey chair. This device was basically a plywood board with a hole that clutched the neck of the monkey so it could not escape. The animals then had to watch simple patterns on a turning disc and were offered pieces of bananas and oranges as a reward. The purpose was to imprint them with one particular color scheme in order to generate a moment of attention in which the corresponding neurons would fire, allowing researchers to establish a connection between the pattern on the screen and the pattern in the brain.

Another research group in the same institute kept more monkeys. These did not live in the spacious zoo-style building in which our animals were lucky to dwell; they had to subsist in the dark basement in tiny desktop cages. They could not really move anyway. They had already been implanted with their electrodes, simple ones but so bulky that their skulls had to remain open. With the measuring apparatus protruding from a huge wound in their head, the animals had to be fixed in their cages with ribbons, like lunatics in a 19ᵗʰ-century madhouse.

When I saw these creatures for the first time in the gloomy hell of the basement, I recoiled and immediately resolved never to go there again. Our approach is much more elegant, I told myself; our animals won't need to endure an open skull. They will only sport an elegant little plug, which will soon vanish beneath their fur when it regrows.

9:14 a.m. We forgot to redispense the Ketanest. I felt the heartbeat of the rabbit. It hummed like a dragonfly in flight, far more frenetic than usual in such a small animal. Curare simply paralyses the muscles without having any effect on the pain. Only the Ketanest anesthetizes the animal and takes away the sensation of agony. But it is late and we had missed the right steps in the anesthetic schedule. If we administer too much Ketanest, we might kill the rabbit. If we do not give it anything, we will be drilling into its head while it senses every motion, unable to move. In only about half an hour, its strength will be back. Then the rabbit will start to thrash and writhe on the operating table, and it will try to get rid of the wires in his head.

What we did on that morning, inside the romantic walls of the old institute, is called *vivisection*. Vivisection still happens here today, 20 years after I assisted in that day's surgery, and is happening right now in thousand places on this planet. Vivisection is still possible because we treat everything that cannot speak our language, as a thing without feelings — plants, animals and, until some decades ago, even children before they reached the age at which they could speak. This custom has a long history. It is the history of our Occidental picture of the world. For thousands of years, this picture has culminated in a rational god representing razor-sharp, unerring and perfect reason, standing outside the world as it is. Only that which came close to this reason, which was in this god's own image, had dignity and merited being protected. The god of reason is immeasurably distant from the animal, which in every moment is totally immersed in its world and whose intelligence cannot be extracted from its body and from the sensual connections this body makes with the remainder of the world. If philosophers had consciously tried to construct the opposite of an animal, they would have come up with an Occidental god of reason. This god is the opposite of the animal in ourselves, of the animal that is *us*.

WHEN DOES FEELING START? AND WHERE?

It is only in the last three or four decades, since the birth of ecological consciousness, that people have started looking for a value system that was applicable not only to rational subjects but to all beings. We have not come very far, though. The new paradigm of ecological ethics from the 1980s has generated a few tenure track professorships, and at least in Europe some of these have already been canceled. It has inspired some hot debates in academic circles, which were relevant only for the initiates. But it has not convinced many people.

A growing cluelessness reigns or, rather, a silent acceptance of the only scale for measuring good and bad that has survived so far: the bio-economic idea of success as outcompeting other contestants. In the absence of a larger, encompassing understanding of what life is, we pragmatically let this mainstream definition reign. In my 2013 book, *Enlivenment*, I try to show to what extent a near-impenetrable compound of evolutionary and economic efficiency thinking has deeply defined all aspects of our world today, impeding us from seeing what life truly is. If we cannot grasp what life is, we will inevitably fail to develop an ethics for it. It is futile to even try; we must understand life first.[1]

Yet we don't. We even systematically deny what we could know, what we could experience through our flesh through the shivers of our neck hairs, which, like mine on that March day in Freiburg 20 years ago, bristle if we are about to damage life. But we are hiding behind a wall of technical knowledge. Mainstream science simply does not have a picture of the organism that is larger than the idea of a functional machine, an assembly of non-living parts.

The inadequacy of the categories that can be derived from the mainstream technical understanding of organisms can be seen in the debate about embryonic and fetal rights. This is an area of huge medical (and commercial) interest. The "mining" of embryonic stem cells has been promising solutions for a host of medical problems that classical reductionist medicine has not been able to solve. But the discussion has been stagnating for years. Human embryos are the gray area where the human rational subject emerges from the silent body of the animal. The attributes tacitly used to discern the human sphere from the remainder of life — "rational," "personal," "useful" — do not apply yet to human embryos. But to most people it is nonetheless clear that human embryos cannot be disposed of or used up as though they were parts of a machine. The fetus in the mother's womb seems to miraculously accomplish the transformation from a mere thing to a subjective individual — a person. But how? And when? And through what means? Researchers and philosophers painfully labor to define the moment when the tiny tadpole in the mother's womb turns into a human subject. Does it happen in the seventh week of pregnancy? In the 13th? When the fetal fingers spread, when the arms start to paddle, when the back bends? At what moment does a human soul enter the animal body? This is the underlying question with which researchers struggle. Its character is still stuck to the Newtonian-style thinking.

Instead of incessantly (and half-heartedly) searching for the legally relevant moment of the beginning of personhood, we should openly accept that this moment never comes — or, always, has already passed. The body does not give any clues. It seamlessly glides into its future form. It is becoming, and it will only become and never be. Life gives birth to life and has been doing so in an uninterrupted chain from the first beginning and, through this, is part and parcel of our own being. We are still directly connected to the ground zero of embodied existence. Being a body, we remain an animal. We have never left the animals' world. Deep inside we know that there is no true frontier where the human separates from the rest of life. Somewhere inside us we are still embryos; we are still the egg just fused with the sperm; we are still a huge cell with a much simpler composition than the slipper animalcule. We still do not know the answer to the enigma of our origin. We still ignore the process "that mingles breath and sense in a way that no one has explained and no one ever will," to express it in the words of the Nobel laureate for literature, J. M. Coetzee.[2]

The crisis of nature and the current dilemma of Western civilization, which some people refer to as the crisis of our values (namely the all-pervasive baseline of efficiency and purely individual utility), are not to be separated. Rather, these form two facets of the same underlying attitude. They both follow from the way we understand living beings, the living biosphere and ourselves in it. This understanding, deeply informed by the "value-free" ideology of the survival of the most efficient, already bears an ethics at its core, even if it pretends not to do so. Its tacit ethics is the law of dead matter and the corresponding fiction that only by adhering to these laws can we, one day, finally step away and arrive at some blissful state where all lack is fulfilled. It is the ethics of pursuing an arms' race to stop wars. It is the ethics of necessary slavery to be free in some distant day in the future. It is the ethics of total surrender to the necessities of the bigger whole to someday gain absolute individual sovereignty.

Our picture of an organism (which always includes how we understand ourselves, as we are also organisms) suffers from a crisis of freedom. The crisis of freedom is related to the fact that we see "nature" and "autonomy" as two distinct parts and not as two sides of a necessary whole incessantly interpenetrating one another. We have to realize that freedom self-organizes within feeling bodies and throughout a meaningful biological history and is not linked to otherworldly reason. Only then, will we understand how deeply values are interwoven with life itself and how little they are the result of a bodiless subjectivity.

9:23 a.m. Ines has asserted herself. With a careful incision, she cuts through the meninges, which, before haltingly giving way, press deeply into the soft mass beneath them. But first we had to wait until the additional dose of Ketanest started to have an effect. The professor took a rough guess at how much of the drug would do, and we injected the liquid under the loose skin of the rabbit's neck. Hopefully the dosage was enough, and we will not have to start again. The rabbit's heart still races like an overheated motor. It does not change its frequency as Ines starts to sink the tiny wire brush at the metal spindle's tip into the brain. The professor stares at the monitors as Ines turns the ratchet with her fingertips. On one of the screens, then on another, a regular pattern appears — nerve impulses. The professor hits his palm with his fist like a soccer coach when his squad scores seconds before the end of the game. The rabbit's heartbeat does not stop racing. We still have 22 minutes until the animal will start to move again.

AN ETHICS SOLELY FOR HUMANS?

Most of the reasons found by humankind today, when speaking in favor of the protection of non human life and for the preservation of nature, rest on one of two traditional polemical pillars. These two versions of arguments for a bioethics are typical of the way our culture overlooks the necessary connection of inward experience and body in one's self and in other individuals. They are not necessarily wrong, but they are lame. Neither of them touches the crucial point that organic being in itself gives rise to sentience and subjectivity and hence cannot be understood by any morals that are applied from a rational standpoint outside the living body.

Let us take a closer look. The first ethical pillar in favor of the preservation and protection of nature and other beings rests on claims for their usefulness. It is about the biosphere's functionality. It calls for preservation because plants and animals, in some respect, are believed to be *our* resources. The most recent offspring of this thinking is the "ecological economy" — the proposal that we should put a price tag on natural resources and hence introduce nature into the marketplace to stop its being exploited for free. Nature, the argument goes, provides ecosystem services, which include clean water and air, pollinators for food crops and beneficial, yet-to-be-discovered drugs. American researchers collaborating with ecological economist Robert Costanza in the 1990s calculated that the biosphere provides goods and services worth 50 trillion dollars a year, with 38 trillion alone attributed to "aesthetic, spiritual or scientific benefits".[3]

Obviously though, in such a picture, scientists have forgotten to factor in death. As living beings, we do not really have a choice in the way consumerism would have us believe. If the biosphere stops delivering its services, we can no longer exist. At the end of all computations, the price of life is endlessly high. If you try to break this endlessness down to the proportions in which single species contribute to it, you run into a paradox. Obviously, if you divide a value that is infinite, the fractions must also be unlimited. The value of the whole of life is immeasurable because it is all there is, because it cannot be broken up into neatly separated parts and because it includes ourselves, and we cannot detach from the scene as long as we are alive. But to be able to calculate ecosystem services, researchers need to put finite price tags on its contributions. This act converts the infinitude of the biosphere into a picture that is objective and can be computed but is still wrong.

The same distortion happens if you try to discern which species hold the keystone of the whole building of a particular ecosystem in place. Although ecologists love to single out "keystone species," any member of a given ecosystem can have a crucial function in tying the web together. If you start to extract members of a given habitat, at some point, which is unknown beforehand, the whole thing crashes. It is held together by the number of relationships in addition to the presence of individual species with particular functions. Here the same logical dilemma occurs: we cannot single out individual contributions. We also cannot untangle them from the effects of our own actions, even less so from our needs. And here we find ourselves in the same situation as the physicists of the early 20th century. They wanted to decide whether light is a wave or a particle and, in the end, they had to admit that it is both. We cannot objectively separate an individual from the whole without deeply misunderstanding each side. Individuals are real. The whole is real. But one exists only through the other.

All this notwithstanding, ecological economists continue to come up with a lot of numbers. But in their computations other living beings do not attain anywhere near the price tags of human products. The value of a single whinchat, for instance, a songbird of meadows and marshes nowadays rare in Europe, has been estimated by ecological-economical calculations at 150 euros.[4] We can easily see that in real life, although the "green accounting" approach convincingly shows that there is no market without nature, it all too easily prioritizes the human dimension and still underprices the contributions of nature.

A slightly varying version of current environmental ethics can be described as the ecological variant of the economical resourcism described above. It takes the biosphere as a resource necessary for human life. Without its complexity, we could not survive. This is the standard, traditional motivation for nature preservation, particularly in policymaking. Nowadays it is included in political programs of all parties and even in many corporations' sustainability statements. Its convincing force goes back, at least, to the legendary report on the limits to growth published by the Club of Rome in 1972. Since then, it is undisputed that disturbing natural complexity has effects that can be dangerous for our survival. Many biologists and environmentalists are deeply convinced of the complexity argument. Often, it is the motor behind their will to protect nature.

But this argument is equally human-centered, and equally centered around functional value. It follows the idea that a certain amount of nature is required for us to thrive. We need it for ourselves. This second pillar sustaining the value of nature no longer creates much controversy. We all want to keep natural complexity, don't we? Right. But as long as we still regard this complexity as a means to our ends, we will remain in a position external to nature, unable to change our relationship towards life in a substantial manner.

The second polemical pillar in favor of the importance of nature explicitly tries to avoid resourcism. It concedes an intrinsic value to all things natural. Here we come to the crucial core dilemma of environmental philosophy. It is easy to speak for the importance of nature as a resource in the current bio-economical framework. But it is nearly impossible to argue for an independent value of living beings. There is no space for this attitude in our present-day understanding of the living. If we take it up, we need to step out of the scientific agenda.

The idea of intrinsic value inevitably collides with the current principles of rational science for which nature is defined as that which is value free and must be described in terms of cause and effect. Here, other living beings cannot have intrinsic values because only man can make judgments about values. Subjective rationality is needed in order to have access to the realm of moral principles, so the argument goes. A body is not enough. I am quite sympathetic with the intrinsic value approach, but I feel that it should attempt an answer as to why intrinsic values come about and how reality generates values in the first place. What is missing is an idea of how real values come into the world on their own accord, and not through human decision. If the term "intrinsic" is to be taken seriously, we need an account of how an interest in oneself can arise on its own. Otherwise, "intrinsic value" ultimately amounts to a moral (or religious) stereotype, as in "all species belong to God's creation." Here, "intrinsic" ultimately is "functional value" as well.

A proxy for the intrinsic value argument is the enlargement of the group of agents who by their subjective nature automatically are moral subjects. If every animal (or plant) had a humanlike consciousness, it automatically would be a moral agent. But normally, the only animals included are those that are highly social, able to recognize themselves as individuals through standardized tests, and in other ways seem to be quite like humans — the great apes, dogs, gray parrots and dolphins. But even this enlargement of the class of rational subjects who can be understood as "persons" in the human sense does not solve our dilemma. For it cannot tear down the wall between "value-free" nature and humanlike agency that mainstream scientific thinking has erected. It cannot answer at what moment an embryo becomes a human being or at which step in the ladder of natural history organisms came to experience personhood. It is impossible to draw a demarcation line through the biosphere, separating the "real subjects" from those beings supposed to be machines. When we try to do this, we always run into areas where a clear distinction cannot be made, just as physicists could not longer clearly decide if light comes in waves or particles.

THE ETHICS OF PITILESS EFFICIENCY

To my eyes, on these grounds we can easily understand why so many approaches to erect an ecological ethics are ultimately not convincing. They tacitly presuppose the separation between special (human) agency and an ultimately inaccessible nature — or they posit an intrinsic value without being able to explain how this comes about, partially conflating "intrinsic" with "functional" in a religious sense.

To exacerbate things, an additional dilemma makes arguing for an organism-centered ecological ethics particularly difficult. This is the general functional undertone of our worldview created by

331

evolutionary theory and neoliberal thinking. The presence of this functionalist bias is often invisible because it is absorbed in the way our culture perceives reality. Evolutionary success, as it is conceived by biologists, does have a moral: an organism needs to be efficient, to outcompete others, to behave like a well-functioning, thoughtless machine with the purpose of propagating its own genes. In this ambiance, it is nearly impossible to devise an ethical theory extending to organisms that escapes the functionalist notions of evolutionary optimization. If biological traits have an overarching selection value (the "currency" ecological relations are mediated with in much biological thinking), it is difficult to see how they can also have an intrinsic value.

It is important to see that the evolutionary paradigm is a mirror image of neoliberal economic theory. Both in fact can be understood as one huge, overarching thought pattern structuring how we perceive reality. I have tried to understand the underlying morals more deeply in my book, *Enlivenment*.[5] Here, I want only to stress that the ideals of utility maximization and efficiency through the "selfish gene" or "homo economicus" already include morals that clearly define the ethical signification of others in relation to self. For these morals, others are the means in an evolutionary theater. Actually, all living individuality becomes a mere function in an optimization game. For the bio-economical ethics that tacitly prevails, even the experience of self is only a means towards the sole purpose of winning the struggle for life.

We should not underestimate the depth to which this bio-economical thinking affects real life. Philosophers in their ivory towers cannot help because they cannot find a convincing connection between a living being and the world it lives in. The whole situation is a clear double-bind: mainstream science denies having moral attitudes while positing a crushing behavioral ethics outside of which no scientific contribution is to be taken seriously. We know from the pioneering work of Gregory Bateson that a double-bind is always a sign of a no-win situation in relating.[6] Developed by Bateson in order to understand the factors contributing to schizophrenia, the term can be generalized to describe any deep attachment pathology in which what is said is contradicted by what is done. Creating double-binds means to deny reality and to force others, even if they see things correctly, into the same denial. This could easily be said about the way industrial civilization relates to the biosphere to which it belongs.

We, therefore, are in dire straits. There is no conventional scientific argument that can support the intrinsic value of natural beings. Our choices are no-win. Either we rely on a rational set of values stemming from identifying other beings as resources and, therefore, are prone to neglect and destroy nature. Or we follow our intuition that there is something deeply important about the presence of other beings and, therefore, we are forced to preserve it. But we then leave the terrain of mainstream scientific discourse, which denies value to anything that is not human and rational, and risk being insulted as dreamers and oddballs. Or we just shrug, declare ourselves unable to understand the living world and, by this, silently hand control to the bio-economical paradigm.

We lack an argument that can serve as a bridge between the world of scientific observation and the realm of subjective value. Only this can be the stepping stone for a comprehensive environmental ethics. The crucial requirement for this type of ethics is that it must show how organisms produce value and how they act according to meaning. If we can argue for the value that arises from biological world-making and needs fulfillment to continue embodied experience, we can create a basic set of existential morals. These can formulate an "ethics of the flesh," an ethics of the body-in-connection. This would be independent of human perspective and its limitations. These are not only ecological in the sense being valid for ecological structures; these also describe actions from the perspective of its ecological signification, which include any productive relating. Only through a concept of values grounded in embodied existence can we search for a natural ethics that does not rely solely on the human decision of what is to be included into the sphere of sentience and subjectivity and what is not.

To arrive here, we need to recall the central observation made throughout this book: subjectivity is coexistent with all living processes and not a reserve of the rational human mind. Life always starts with an individual perspective that brings value into the world. There is the faint voice of the core self, which judges what is individually good and bad in every utterance of life. Environmental ethics needs to hear this voice. We need to follow the delicate track it lays out to imagine the ecological principles necessary to allow for a productive process of living-self-realization in connection with others. Every living ecosystem can show us the way. All this time, it has enacted exactly what we have sought. It establishes a balance between the individual and the whole, allowing both to continue.

9:42 a.m. Ines turns the micrometer to push the ratchet forward. Her eyes dart back and forth from the skull to the oscilloscope. The professor has switched on the loudspeakers. A nondescript swooshing sound fills the room, interspersed with the rattle of the neurons firing. The voice of life, as science believes. Six monitors are filled with decent-looking nerve patterns. The result appears far better than expected. Then suddenly everything disappears; the screens only show gray snow. "What is going on?" the professor asks. "I don't know," Ines says. She turns the ratchet back and forth. Swoosh. An ordered picture shortly appears on one of the screens, then nothing. Where in the brain are we, anyway? In the visual center, as planned? Are the neurons, which we have touched, still alive? The professor pushes Ines's hand away from the ratchet of the stereometric apparatus. Ines twitches, her finger is caught in the electrode wire; the brusque movement tears off the holder. Hiss and crackle. The professor yells, fiddles hectically at the setup. Both are sweating. Four hands desperately try to put the spindle back into position. The careful adjustment we had done at the beginning is now worthless. At any moment, the curare can cease to be effective. The professor screams, Ines yells back, then she storms out of the lab and slams the door. The professor pushes the electrode back into the soft brain tissue, jiggles forward and back, until some of the monitors show signs of neural action. There is a dull knock when the right hind limb of the rabbit hits the alloy table. Then again and again. "Quickly!" the professor shouts. I mix the dental cement with water. The powder is not thoroughly moistened and clots form. The professor sloshes a spatula-load on top of the wound with the plug at its center. The animal is jerking. I press its head down on the steel table, while the professor tries to paste the cement into the hole. Ines has returned and watches us pensively.

ORGANISMS CREATE AN EXISTENTIAL ETHICS

Nearly all Occidental value systems have the same weak spot. They put the idea of morals first and the experience of reality second. They do not waste thought on how a moral subject is related to its world. Or how it feels. Occidental ethics is top-down, but we need a bottom-up approach. We need an ethics that follows from the relationships that sentient bodies have with other sentient bodies. A "wild ethics" as environmental philosopher David Abram would have it. An ethics able to make us "think like a mountain," as Aldo Leopold, who coined the term "land ethics," postulated. An ethics that is not restricted to moral subjects but can extend to what is good or bad for embodied subjects and the relational systems they form. An ethics that becomes visible through the way interrelated beings thrive. An ethics incorporating beauty as an intrinsic scale of success. To be able to imagine this sort of ethics, we must remember that the feeling body is the ground zero of any morality, the origin of everything good and bad.

However, morality, as we know it, always has had the tendency to neglect the real being of subjective agents. For the prevailing Occidental ethical concepts, man above all is a rational, but not embodied, being. The dimension of nature, of body, of sentience, which is the prerequisite for any rational thought, is rarely included. This neglect has shaped most of our philosophical approaches towards ethics, starting in the age of classical Greek antiquity, where the idea of pure rationality celebrated its first major breakthroughs. Plato, for instance, dreamt of a world governed by eternal truth. For him, the living figures we encounter in the world and whose skin is bathed in the sun's light were, in reality, nothing but shadows — trembling mirages confusing our senses. In his famous cave parable, Plato compared living things to the jittering dance of shadows a prisoner sees when somebody passes along a small hole in the otherwise closed entrance of a cave. What we can grasp of reality, so the message was, is distorted and brute. Real reality is beyond, in a world made not of flesh but of ideas.

Plato's attitude has deeply informed the Western viewpoint; the "true" world is imagined to lie outside of this one and, therefore, true morals can only be gained from beyond. They cannot be derived from the shadow world we inhabit. The only thing we can do down here is to coerce the world of bodies to more resemble the world of ideals. It is easy to see that the theoretical edifices of rational ethics follow from the Platonic heritage and even the invisible fourth pillar of bio-economic ideology does so as well. Declaring subjective experience a mere means in the struggle for survival is substantially congruent with the manner in which Plato outclassed real, bodily experience with something that "in reality" is pure, but is no longer accessible to human experience. The hardcore evolutionist version of the cave parable is Richard Dawkins' story of the selfish gene: we are all telecommanded by the blind selfishness of our DNA, which only uses the body and all its experiences to maximize its own multiplication. In this view, experience really has become a shadow.

I cannot stress strongly enough the degree to which this other-worldliness has shaped our Western relationship with reality. Morals were thought to be the link connecting us with an otherworldly god, thus leading us from the dungeon of Earth. One of the last great architects of a pure rational world was the Prussian thinker Immanuel Kant. As a person, he was gentle, albeit tortured by neuroses and phobias all his life — he was not able to travel, giving up twice an attempt to leave the town of Königsberg as soon as he was outside the city walls. Kant's moral philosophy was conceived with the iron strength of a Prussian army travel order. His "ethics of duty" did not necessarily mean that the moral subject prospered while obeying the ethical laws; most important was that they did it right, that they fulfilled their moral obligations. The philosopher Jürgen Habermas demonstrated he was still thinking in the same abstract vein when he constructed his discourse ethics in the 1980s. Habermas proposed to determine values by a committee of ideal discourse partners, which the philosopher imagined to be perfectly rational. They had to be disembodied subjects in order to derive the right moral decisions.

But we are earthly beings, weak of flesh, and our bliss consists exactly of this weakness and earthliness. We do not need an ethics for perfect discourse partners because these do not exist. We need an ethics for irrational subjects. We need an ethics for an incomplete creation. We need an ethics that is imperfect from the beginning, provisional, an ongoing tinkering work itself, and which, for exactly these reasons, can be applied to our lives in a living world. Ethical judgment is not adequate for an embodied subject whose mere biological life process excludes the achievement of perfection but is about negotiating the best compromise between individuality and the whole. The goal of an ethics, which is truly beneficial for life, cannot be control but must be healing.

In a number of remarkable books, primate researcher Frans de Waal demonstrated that morals do not fall from the sky. Basic moral reasoning or rather social emotions, as de Waal calls the precursors of human ethics in the animal realm, are not exclusively reserved for humans. Chimps and bonobos, for example, have a system of social guidelines governing mating behavior. In order to exist in their social group, the animals must obey them; however, this often means some sacrifice and suffering, so the apes try to secretly evade rules if possible. Here the basic type of human moral dilemmas is born. These animals follow the biological laws regulating the coherence of their population. But in doing so, other primates do not behave as automata but rather experience the values of their actions in an emotional form. As we all know from visiting the zoo as children, these emotions are easily visible to human observers if we only watch attentively enough. Apes are not social machines but social agents. They have freedom to decide. This freedom is experienced in an emotional manner. Again, we can see that emotions are a way in which an existential constellation is translated into subjective meaning. Human ethics start here, with the experience of the value of certain social behaviors. "Human morality," de Waal says, "is firmly anchored in the social emotions, with empathy at its core."[7]

Our traditional ethical thinking is rooted in the social emotions of our ancestors. It does not mean that all morals are biologically hardwired or that all biological morals are unequivocally good; it only means that moral feelings have a bodily base that cannot be easily circumnavigated. Moral feelings that arise from the social emotions of gregarious primates are not Darwinian instincts. They rather show that we, as well as other animals, experience the consequences of social actions through feeling and that these feelings immediately establish central values no ethical theory can push aside. The social emotions of animals prove that feelings are the dimension through which we experience reality and ourselves inside of it, including the meaning of social attachments. No ethical reasoning, therefore, makes sense without respecting this emotional reality.

The experience of existential meaning, an active interest in oneself, and hence intrinsic value, do not presuppose a rational subject. But these dimensions unfold for a feeling subject. Every organism is living in a universe of values and, through its embodied core self, it knows what is good and what is bad for its continuation. Embodied existence, therefore, is the primary benchmark for any ethics.

STEPPING BEYOND THE NATURALISTIC FALLACY

However, many philosophers today hesitate to see humans clearly as what we are — as animals, members of a primate species with corresponding social needs, and with their particular, species-specific version of the biocentric tragedy brought about by the necessary tension between the individual and the system it is immersed in. The resulting image of our nature and needs is, therefore, unrealistic; something is omitted. We still ignore a vital part of ourselves, a crucial dimension that grants us the need and ability to connect with everything else.

This ignorance most bluntly manifests in the fact that we have lost our once moderate balance in relating to the world. As opposed to primate societies (which for better or worse manage to live in a certain equilibrium with their tropical forest biotopes) and to early human civilizations, we have forgotten the rules by which we can live in symmetry with the rest of creation. The continued destruction of other life is a declaration of the failure of our present ethical system. Ironically, our ecological predicament could only have come through the deep ethical misunderstanding of ourselves as a species detached from all other life forms.

For ethical values not to become toxic, they must not misjudge the essence from which they were created; they need to respect who or what a moral agent is. Before we start with ethics, we need a functioning ontology: an idea of what reality is about. The possibility of imagining this, however, has been largely washed away by a surge of general suspicion of ourselves. It is fashionable to declare it impossible to see clearly what is. But we often underestimate our natural powers of relating. And we exaggerate our powers of control, which in truth we do not have. We still largely live in an illusion about ourselves.

We are a machine-like animal, so the doctrine says, ruled by unconscious psychological drives, and everything we can experience is limited by arbitrary laws of language. This multiple discouragement is a result of the thinking of the last one-and-a-half centuries, during which we have thoroughly renounced our feeling of aliveness as a tool for understanding the world, relying solely on assumedly rational faculties. One of the outcomes is a deep melancholy — and no wonder. If we convince ourselves that we are something we are not, then we have lost ourselves.

In principle, it should be easy to shed these false beliefs. But in reality, our incarnate confidence needs a lot of practice. As a first-aid kit we should always remember the basics: we are organisms; therefore, we know what is good for organisms. We feel and we can be united with other feeling beings. We live as agents of imaginative change and are allowed to take imagination seriously.

The most general and widespread argument against the idea of natural values and an embodied ecological ethics I am pursuing here has been the blanket prohibition of the naturalistic fallacy. In philosophical circles, you are blameworthy of deadly heresy number one if you infer what "ought" to be from what "is." This prohibition lies at the ground of our current science's obsession with a "value-free" description of reality: only human agents can attribute values.

If we look at what organisms continually do — inferring what ought to be for them in order to be in the future — we can see, however, that living beings constantly commit the naturalistic fallacy. They continuously infer what has to be done from their momentary being. The body is the paradigm of what *is* and what *ought to be*. An organism's body has needs because it is made in a specific way. Through being what it is, a body has needs that tell it what ought to be done. As the world is incarnate and real only in the dispersed bodies of sentient beings, "is" and "ought" always intermingle.

Organisms represent matter in a specific arrangement whose particular feature is that it desires to persist over time. Their essence lies in what ought to be done in order to exist in the next moment. An organism is defined by its desire to be. Here, we can see the "ought" which it creates through its sheer being, in a clear shape. We can here see again the quantum leap that a poetic ecology tries to take: the gap between a rational subject and law-like object in the Newtonian view now closes to bring forth one and the same subject-object. We, the organisms, always are both the observer and observed. We are world that simultaneously watches and feels itself. This is the most basic principle for any ecological ethics. It, therefore, does not make any sense to prohibit an inference from "is" to "ought", if one's own body is doing it all the time.

Philosophers are hiding behind the fear to commit a naturalistic fallacy and prefer not to make any positive ethical statements. Bio-economics, the compound of evolutionary thinking and liberal economy we discussed some paragraphs above, ironically does not know such fears. Bio-economical thinking is invisibly influencing the image we have of ourselves and of the world, although it rarely explicitly pretends to offer ethical advice. Its moral impacts are conveyed in a rather subtle way as the paradigm invades our entire conception of reality.

Ironically, the leading mainstream idea of reality is itself based on an inference from "is" to "ought". It is built on a huge, toxic double-bind. Our basic metaphysics does not obey the prohibition that is one of its own central dogmas. The mainstream view holds that living beings are machine-like agents pursuing their egoistic success through competition. Therefore, so the (albeit hidden) conclusion goes, egoistic pursuit of success and competition are the way the world works and the way we *ought* to behave in order to cope. The inference from "is" to "ought", while being prohibited for thinkers who try to understand organisms beyond mechanical thinking, has become the ethical root-metaphor structuring our thought without us even realizing it. But whereas the body is always right in setting its own existential values by existential needs, the hidden mainstream way of inferring from a supposed nature to seemingly inevitable values is not, as felt experience has no place in it.

10:04 a.m. "Catgut, please." The professor squeezes the words through his tightly clamped lips. We are sowing with cords from animal intestines, as though working in the operating suite of a real hospital. The thread ties the two rims of the scalp together, forming a long, wrinkled bulge. The plug sticks out from the bloody construction site like a misplaced Lego block. My fingers hurt from holding the clamps and from pressing the rabbit's shaking head down on the table. The right hind limb of the animal jerks in a mechanical beat like a faulty machine. Then with a short jolt we pull the tube from the animal's gullet. Does it breathe? Or has all this torture been in vain? The rabbit beats its head upward and then lets it slap down. Foam hangs between its whiskers. "You take the anesthesia guard," the professor says to me as he walks out of the room. Ines silently clears away the instruments. We put the animal in a box strewn with sawdust. The rabbit's breath is weak but regular. Its eyes are half open, its gaze sweeping past us, unresisting.

I am now alone with the animal. I take the book I have brought with me from my backpack. It is the famous psychoanalytical biography of Goethe by Kurt Eissler, one of Sigmund Freud's top students in Vienna, who later emigrated to America. I have started reading it resolved not to believe a word of how the author dissects Goethe's unconscious neurotic motives. I still consider myself to be my own master, the master of my emotions and my body. But Eissler's evidence is too good. Against my will I read on and forget the animal for a long time.

When I look up, my glance locks with the black eyes of the rabbit. The animal has sat up. Its nostrils are twitching in the typical way of its kind. It seems as though returned from a great distance. It ignores the fact that it has changed forever, that it is marked for life. On its head, the bloody comb stands out, but for the rest it seems to behave in a totally normal way. I pour a handful of food pellets into a plastic dish and go home. As I make my way through the basement exit, it occurs to me that we never gave the bunny a name.

March 7, 9:04 a.m. The rabbit is dead. It died in its box during the night. We don't know why. The professor fixes a new surgery date for the next morning. We still have the other animal.

CHAPTER 14:

Enlivenment:
Ecological Morals as
Mutuality in Beauty

Nature, life and beauty cannot be untangled.

— Sandra Lubarsky

The Grunewald forest stretches from the western fringes of the inner city of Berlin far south to the urban borders. The wood extends for about six miles until it meets the Wannsee Lake, where in early summer the nightingales sing, those sweet Berlin spring birds we visited in Chapter 8. I often ride my bike through those Grunewald forest areas, which are a bit closer to the city, in order to take my energetic poodle out into the woods. As both dog and owner have become a bit conservative over the years, we usually follow the same route. It leads to an impressive old oak tree, which stands some yards apart from the path winding through the forest. The tree is our destination, a place to rest for some minutes in the presence of a very particular being — or maybe, rather, a whole universe of beings contained in itself, always changing and still always the same.

The tree may be 500 years old. It looks old — battered, rugged, huge. Its bulky trunk has split into several heavy branches, which are partly broken, half alive and half dead. The lifeless arms stand out naked in the air among other twigs aflutter with leaves. A huge bulb grows out of the lower stem, obviously some deformation caused by a parasite, which must also be hundreds of years old. The tree is punctured with holes, where in spring tits and woodpeckers nest, where bats find cover during the summer nights and where hornets build their lairs. From other openings the soft yellow flesh of huge fungi spreads during the growing season. A couple of mistletoes have clutched the branches, and in the earth piling up in some forks between branches, ferns have started to grow. It feels good to arrive here and to rest for some moments in the perimeter of the old living being, which so effortlessly integrates all the contradictions of being alive within its scope, embracing birth and death simultaneously.

My old live oak in the Grunewald forest is an image of an environmental ethics that works and is productive without using any words or other descriptions. It does not follow an abstract concept but manages to integrate a host of diverging embodied interests. They all can coexist if none takes up too much space, if all enter into some sort of mutuality. We can experience that this structure of shared lifelines is successful because it attracts us. It is beautiful. We feel at home here. In the last chapter we have seen that it is impossible to reason about an ecological ethics, or an ethical position concerning the living in general, without taking into account the true needs of living beings. Ethical behavior is a performance that fulfills exactly these needs. In this short chapter I want to show that these needs are always mingled with the needs of others and that all thinking about a possible ethics has to start from this vantage point. This has another crucial consequence: we cannot think about what a good life is if we do not ask ourselves how a healthy ecosystem works.

We need to approach ethics from the ecological stance of massively distributed interbeing for another reason as well. As we have seen, a living being is deeply paradoxical in nature. It is form through matter; it is an immaterial inwardness ruling a body, and it is self through the other. If we isolate subjects from these dialectical exchange processes, the only means through which they realize themselves, then we cut off this necessary paradox. In the living processes of the biosphere these different dimensions are integrated in the same way as the live oak has integrated its own dying into the ever-renewing vortex of diversity it offers to the Grunewald forest.

The leading question for an ecological ethics, therefore, cannot be focused on the behavior of a moral subject alone. It needs to encompass all subjects *and* the whole they are bound together in. Therefore, the problem could be formulated as this: what is needed in order to allow that an embodied subject is able to unfold itself in interbeing

344

with others and in order to let the others thrive through the well-being of the embodied subject?

POETIC MORALS

Let us listen to the living body once again. It follows its values in the most visible, palpable, seizable manner — all contrary to the belief of conventional biology that all strivings of living beings are only semblance, the "just as" of an automaton. Life wants to live on, wants more of life, wants to expand, to swell and to blossom; wants to propagate itself and rise again in a thousandfold manners. Life wants to be subject in an emphatic way. This is one of its two sides: the desire for autonomy. The other side is its need for what it is not: for the matter that is the sole means through which it can develop its identity. This other side is the want for the presence of other beings in which the subject recognizes itself, a yearning for those others, which it can love in order to grow. If the other brings forth myself, then she is an absolute value for me, which stands for itself and does not only arise through rational reasoning. All autonomy is born through the other. Every subject is not sovereign but rather an intersubject — a self-creating pattern in an unfathomable meshwork of longings, repulsions and dependencies.

We hit a paradox here again. If there remains no fixed structure of a being's self as soon as we really proceed into its depths, into the abyss of a "selfless self," then its actual well-being is in effect a gift from the other. We come into being only through the other. Self and other are so intimately interwoven that, if we insist on first discerning what a living being is and does, before talking about norms, we immediately run into its entanglement with other. Other is first. Without other, there is no self. We need the gracious gaze of the alien black eyes if we want to learn to know ourselves. But the primacy of the other for self does not mean that self is not important. To the contrary — self is still the other for an other. Self is part of interbeing and part of an unfolding world. The true locus of value, therefore, can only be the living meshwork. The web. Life that accomplishes itself is that which enhances all the possible relations in this web, which does not cut off one of them but rather reinforces the existing ones and weaves in new ones. We are nothing without nature. We are opposed to it as we are contained in it. We are a fold in its infinite tissue. We are its imagination just as it is ours.

Before we can debate a new ethics, therefore, we humans, the speaking subjects, first need to understand ourselves anew through our symbiotic entanglement with all the other beings. Moral reasoning becomes a question of the language used. Rather than being a rational means to codify objective relationships (or a totally detached self-referential game), language can be our medium for partaking in a larger organic whole. We should stop viewing it as the sharp blade severing us from the rest of nature and rather understand it, as the poet and philosopher Gary Snyder so eloquently argues, as our way to be part of the wild. Language is what welds us together with the silent realm of meanings, for it exposes that feeling which the body only can show in mute joy or suffering.

Let us not forget: it was Orpheus, the singer, who made the trees shiver and yearn to listen to his song. Every human has the ability to speak with the Orphic voice, that poetic way of saying that early Greek mythology remembered from time immemorial. The Orphic voice does not blindly posit, but speaks through listening. It sees with words coming from the inside of poetic space. It lends a voice to the phenomena themselves, which makes plants and stones answer and grants the poet the powers to speak as a part of the world around him. The Orphic voice always comes with existential morals. It has an ethical dimension from the beginning because it invites every being to partake, to speak out in the grand concerto. Poets of all ages have tried to express this common space of meanings that at the same time are inward and outside, both body and sense. These meanings are understood not by finding new expressions for them but are bestowed. "The highest would be to realize that all 'matters of fact' are really theory. The blue of the heavens reveals to us the fundamental

law of chromatics," Goethe claimed. "Let man seek nothing behind the phenomena, for they themselves are the doctrine."[1]
The American poet Wallace Stevens, who in his main job was
an insurance company executive, proclaimed a similar purpose
for writing as "becoming what surrounds me." Poetic expression
resonates with ecological meanings, with the core experience of
being alive, which cannot be exhausted with words. It is ecological, and in being ecological, it contains an ethics of aliveness.

Any ethics must start on ecological grounds. And in being
ecological, it cannot do other than be poetic. It has to achieve
the everyday magic of life in which inwardness and outwardness
become mutually expressive. This entanglement always enacts an
ethics. To speak and to create new symbols by syllables — sen-
tences, sounds, images, gestures — is one of our deep possibilities
for partaking in creative nature. It is the human manner of finding
expression. This is not something that sets us apart from other
living beings but rather is the extension of their natural autonomy, their natural freedom into the sphere of discursive reason,
which is nothing without them.

The speaking subject already speaks for something else. This
other speaks only in order to be heard through the subject. Both are
part of a network of mutual transformations. The ethics we are looking for, therefore, must satisfy two needs: it must first consider how
biological subjectivity comes forth and to what degree the body's
needs are the foundation of all value. Second, however, it must take
into account that any ethics must be equally based on what is good
for the whole — the ecological network — as well as for the subject.
Our ethics, therefore, needs to be massively distributed and always
able to creatively change the subjects for which it manifests. Briefly,
it needs to be rhizomatic. It must, in other words, understand reality
as the web of contiguous transformations the French thinkers Gilles
Deleuze and Félix Guattari have called a rhizome.[2]

A rhizome is real and abstract, felt and fabricated; it is bodily touch, imagination and memory all in one. The rhizome is our

rootedness in the others, our birth through their gestures and of theirs through us. Think of the real plants in the forest and their symbiosis with the invisible fungi in the soil. This meshwork of sub-terranean threads travels through the earth and enables plants to emerge and mushrooms to bud from it. Our language is like a fungal body emerging from this invisible deeper connection, bringing the fruits of a deeper interconnectedness to maturity.

From these deliberations we can sketch the following skeleton of biocentric values. They all modify one principal idea, which is that we need to preserve nature because it is ourselves, and because, par-adoxically, it is everything that we are not.

1. We need to preserve nature as the embodiment of individual needs. These needs are a real "ought," perceived and expressed by subjects. Values have a material dimension. They are self-organizing.

2. We need to preserve nature as the visible shape of inwardness, as reference for feeling, as the gestalt of our own psyche. Without nature, we risk losing important parts of our scope of feelings. We risk losing our ability to love.

3. We need to preserve nature because without it we are speechless. We need plants and animals as parts of ourselves, both in physiological and in psychological terms.

4. We need to preserve nature because it shows us possibilities of existence we could never know alone. These comprise following a principle of plenitude, desiring a maximum of possibilities, seeking continuous development, craving authentic expression, respecting silence, being energetic in doing and devoted in resting, alternating each summer with a winter, each day with a night.

5. We need to preserve nature, for it is the place
   of the absolute other. This absoluteness, which
   some philosophers have called *countenance*, is
   the absoluteness of being. The countenance
   of the real other is the door leading into the
   absolute other. We must preserve plants and
   animals in order to enable them to witness.

This last point is important. A poetic ethics does not strive for perfection but cares to leave open the crack through which suffering, but also the necessary light comes in.[3] Nature is not the all-embracing wholeness that grants salvation. This wholeness, therefore, cannot be the aim of an ecological ethics. Nature is fragile and fragmentary to the core, and only through this fragility can it be creative and life-giving. The values of the living are at once paradoxical and contradictory. This imperfection is the necessary prerequisite for creation to emerge. An ecological ethics must, therefore, center on this imperfection. It is an ethics of mutual accommodation, rather than one of control.

Nature does not confer salvation but healing. Healing means to transform the oscillating dance on the razor's edge of aliveness into the beauty of a new imagination of what life can mean. It is a process, not a state, and thus never to be secured. It is a dynamic balance tied to the moment and to the situation. Healing means to overcome the cleft between the individual and the other, between the individual and the whole, for one short moment. The great psychologist Erich Fromm saw our developmental goal as the union between freedom and relatedness. This union, however, is nothing that can be achieved. It is a contradiction in itself and, therefore, always means negotiation, a solution that is not exhaustive but rather a momentary compromise. Healing does not signify finding the definite answer but responding with another, more interesting question.

LIFE AS ETHICAL PRACTICE: THE
ECOLOGY OF THE COMMONS

An ecological ethics is about finding a way to enable healing. Any creation is imperfect; the enigma of autonomy-in-connection is always painfully real. How shall the compromise be created? What does a balance look like? Here, it is beautiful and very helpful that we can direct our gaze not only to nature itself but also to the many ways in which other civilizations have inserted themselves into ecological systems and have tried to treat nature in a way that both humans and other beings were connected in a continuous process of being mutually healed. In many respects the rituals of archaic cultures and still-existing ethnic groups are about a healing mediation between humans and the organic web. In the cultural symbolic system of these peoples, these ritual powers also grant human existence. The rules leading to this mediation are general, as they are always about life's duality of individual and whole, and they are local, as they enact the bigger picture in a unique way. This is the situation we find in every ecosystem.

In a temperate forest like the Grunewald, which has given rise to my beloved live oak, there are different rules for flourishing than in a dry desert. Each ecosystem is the product of many rules, interactions and streams of matter, which share common principles but are locally exclusive. This strict locality follows from the fact that living beings do not only *use* the ecological meshwork provided by nature; they are physically and relationally *a part of it*. The individuals' existence is inextricably linked to the existence of the overarching system. The quality of this system, its health and beauty, are based on a precarious balance, which has to be negotiated from moment to moment. It is a balance between too much autonomy of the individual and too much pressure for necessity exerted by the system. Flourishing ecosystems historically have developed a host of patterns of balance, which lead to extraordinary refinement and high levels of aesthetic beauty. Hence, the forms and beings of nature can be experienced as solutions that maintain a delicate balance in a complex society. These solutions are functional for life, and in being so they make our core self resonate. Therefore, the embodied solutions of individual-existence-in-connection exert that special beauty of the living, which fills most humans with an experience of meaning and belonging.

Nature as such is the paradigm of distributed subjects, which realize themselves only in mutual transformation. Any ethics, therefore, must start by taking this extreme form of mutuality seriously. Nothing in nature is subject to monopoly; everything is open source. The quintessence of the organic realm is not the selfish gene but the source code of genetic information lying open to all. As there is no property in nature, there is no waste. All waste byproducts are food. Every individual at death offers itself as a gift to be feasted upon by others, in the same way it has received its existence by the gift of sunlight. There is a still largely unexplored connection between giving and taking in which loss is the precondition for productivity.

In the ecological commons a multitude of different individuals and diverse species stand in various relations with one another — competition and cooperation, partnership and predation, productivity and destruction. All these relations, however, follow one higher law: over the long run only behavior that allows for productivity of the whole ecosystem and that does not interrupt its self-production is amplified. The individual can realize itself only if the whole can realize itself. Ecological freedom obeys this form of necessity. The deeper the connections in the system become, the more creative niches it will afford for its individual members.

As we can see, a thorough analysis of ecology can yield a powerful ethics. But this ethics is more than a set of principles of moral actions towards other beings. An ethics that accepts being deeply entangled with the self-producing values of living beings is at the same time a view about reality, an ontology. The ethics we are looking for binds together an understanding of reality, the principles for remaining in interbeing with that reality and, as the most important guideline, that reality is above all a process of self-creating freedom. In respect to these requirements we can see that natural processes can define a blueprint to transform our treatment of the embodied, material aspect of our existence into a culture of being alive. Ecological creativity, which includes humans in a larger metabolism, provides the binding element between the natural and the social or cultural worlds. To understand nature in its genuine quality as a system of ecological transformations opens the way to a novel understanding of ourselves, in our biological as well as in our social life.

Although the deliberations that have led us to this point stem from a thorough analysis of biology, their results are not biologistic but rather the opposite. Our analysis has revealed that the organic realm is the paradigm for the evolution of freedom. The necessities resulting from that basic principle are non-deterministic. They are rather grounded in an intricate understanding of embodied freedom and its relationship to the whole: the individual receives her options of self-realization through the prospering of the life/social systems she belongs to. To organize a community (between humans and/or nonhuman agents) according to the principles of embodied ecology, therefore always means to increase individual freedom by enlarging the community's freedom (see the table below).

Implicit Ethics in Different Views of Nature

Darwinism	Enlivenment
DISPLACEMENT	TRANSFORMATION
RESOURCE DEPENDENCY	DEPENDENT FREEDOM
SEQUENTIAL OPTIMIZATION	INTEGRATION
SURVIVORS	SUBJECT-IN-COMMUNITY
LOCAL	LOCAL AND GLOBAL (HOLISTICALLY INTEGRATED)
SUSTAINABILITY = EFFICIENCY	SUSTAINABILITY = FELT MEANING
PREDATION AND DEFENSE	OPEN SOURCE
WINNER TRANSMITS MOST GENES	WINNER MORE DEEPLY INTERWOVEN WITH COMMUNITY
EFFICIENCY	DIVERSITY OF EXPRESSIONS
DOMINANCE	SHARING
SPECIES UNDER SELECTION PRESSURE	COMMONS
SEPARATION	PARTICIPATION

Contrary to what our dualistic culture supposes, reality is not divided into separate domains of matter (biophysics, deterministic approach) and culture/society (non-matter, nondeterministic or mental/culturalistic approach). Living reality rather depends on a precarious balance between autonomy and relatedness on all its levels. It is a creative process, which produces rules for an increase of the whole through the self-realization of each of its members. These rules are different for each time and each place, but we find them everywhere life is. They are valid for autopoiesis, the autocreation of the organic forms but also for a well-achieved human relationship, for a prospering ecosystem as well as for an economy in harmony with the biospheric household.

The ethics we need to look for in the realm of living things, therefore, cannot be a set of abstract principles. It must be a practice of realizing oneself through connection with others, who are also free to realize themselves. Gary Snyder calls this a "practice of the wild." If we look to the ways other cultures have tried to become a creative part of ecosystems, hence to actually practice the wild, we can observe that the form they do this is what we would call a commons. The other beings are not an outside nor a resource. They share a common productive and poetic reality.

Historically, we understand by "commons" an economic system in which various participants use the same resource and follow particular rules in order not to overexploit it. If we look deeper into actual commons principles, we can see that the traditional commoners do not distinguish between the resource they protect and themselves, as users of the resource. The members of a commons are not conceptually detached from the space they are acting in. The commons and the commoners are the same. This is basically the situation in an ecosystem.[4]

The idea of the commons thus provides a unifying principle that dissolves the supposed opposition between nature and society/culture. It cancels the separation of the ecological and the social. In any existence that commits itself to the commons, the task we must face is to realize the well-being of the individual while not risking a decrease of the surrounding and encompassing whole. If nature actually *is* a commons, it follows that the only possible way to formulate a working ecological ethics — which inserts the human right in the middle of nature and at the same time allows for freedom of self-expression and technological invention — will be as an ecology of the commons. The self-realization of *Homo sapiens* can be best achieved in a system of common goods because such a culture (and thus any household or market system) is the species-specific realization of our own particular embodiment of being alive within a common system of other living subjects. The commons philosopher and activist David Bollier claims accordingly: "We need to recover a world in which we all receive *gifts* and we all have *duties*."[5]

ETHICS AS FIRST-PERSON ECOLOGY

Agency is always inscribed within a living system of other animate forces, each of which is both sovereign and interdependent at the same time. In the commons, humankind does not hold arbitrary sway as a ruler but plays a role as an attentive subject in a network of relationships. The effects of interactions reflect back on those acting, while all other nodes, animated or abstract — human subjects, bats, fungi, bacteria, aesthetic obsessions, infections or guiding concepts — are active as well. Every commons, therefore, can also be described as a rhizome — a material and informal network of living, incarnate and meaningful connections, which constantly changes as it mutates and evolves.

The innermost core of aliveness cannot be classified and negotiated rationally. It is only possible through being involved in experiences and creative expression. That is why the idea of the commons, which is fundamentally about real subjects seeking nourishment and meaning through physical, pragmatic, material and symbolic means, is the best way to describe an ecological connection to the rest of the biosphere and to provide a blueprint for an ecological ethics. For a commons is always an embodied, material, perceptible, existential and symbolic negotiation of individual existence through the other and the whole. It is an attempt to echo the forms of order implied in the self-creating wild through acts of creative transformation in response to the existential imperatives of the wild.

This dimension of living reality, therefore, should follow a *dialogic* rather than a binary logic, as French philosopher Edgar Morin claims. Morin's dialogic does not try to eliminate contradictions but explicitly thrives on them. It is a logic of dialogue and polyphony;of encounters, conversations, mutual transformations and interpretations; a logic of negotiation and striking compromises. It is this stance of negotiating, adapting and enduring that has determined the way in which humans have dealt with the more-than-human world since time immemorial.[6]

An ecosystem through its shape as commons not only integrates agents and the whole, which these agents build up. Its reality is at the same time material and structural, experienced and created. It, therefore, combines subjective and objective perspectives. Emotional experience is not alien to the conception of an ecological commons but central to it. In an ethics of mutual ecological transformation, feeling is a central part. As inwardness is the necessary way bodies experience themselves, feeling is also a crucial component of an ecological ethics. It is not an add-on that might be tolerated; it is inextricably linked to the reality of ecological functioning. If a living being participates in the exchange processes of an ecosystem, it also gets emotionally involved. This emotional dimension is how living beings experience the relevance of their connections, the meaning of how others reciprocate and how the whole setting acts on their self-productive process. To be connected, to be in metabolism, is always an existential engagement, and this echoes as feeling. Feeling is, so to speak, the core self of a commons ethic. It symbolizes how well the mutual realization of individuality and the whole are achieved.

Indian geographer Neera Singh has shown the extent to which this emotive power encourages commoners to act and provides subjective rewards for their action. She demonstrates that villagers in rural India not only make resources more productive through their commoning with forests. They also satisfy emotional needs and "transform their individual and collective subjectivities."[7] They are engaging in an active poetics of relating in which the human affect and the material world commune with each other and alter one another, in which inwardness is expressed through living bodies and material objects always have a symbolic and felt aspect. Participating in a commons of this kind for a human means to fully realize her ecological potential and to experience this realization through the feeling of living a full life. Again, as I pointed out in Chapter 9, this constellation is known by a common term: we call it love.[8]

BEAUTY IS HEALING

Our capability as living beings to inwardly experience the existential meaning of outward relationships gives us a means of emotional ethical evaluation. We always automatically assess the degree to which an ecosystem, or any relational structure we are involved with, is able to grant us the freedom to be and to be in connection. This evaluation is part of the process of living and hence of relating. Inwardly, this is the feeling of being alive, the experienced aliveness. Feeling alive or "enlivened" is, therefore, an immediate way to experience whether a set of relationships is healthy or not. We feel what J.M. Coetzee described as joy, as the experience of full living. We could also call it the experience of beauty. It is an experience which connects the perspectives of first and third person, the observation and the felt meaning.[9]

Therefore, "where there is much life, there is the potential for great beauty," as the American environmental philosopher Sandra Lubarsky observes. Beauty "is not a quality — blue or shiny or well-proportioned or a composite of these — overlaid on a substance. It is not owned by the world of art or fashion or cosmetics. ... It is embedded in life, part of the dynamic, relational structure of the world created by the concert of living beings. And it is what we name those relational structures that encourage freshness and zest so that life can continue to make life. ... Life, wilderness, biodiversity, and beauty are an interlaced knot; when the cord is cut, the intricacies are lost, the entire weave undone."[10]

By the experience of beauty we are able to evaluate the life-giving potential of a situation or an ecosystem. Beauty, therefore, as a sign of an enlivening situation, is itself giving life. Any aesthetic experience of nature thus is to some degree an ethical assessment. Ugliness, on the other hand, has a certain degree of toxicity. The functional desert of contemporary agricultural landscapes with its few species leaves us uninterested, whereas the Mediterranean dry slope with its rose bushes and bluebirds makes our hearts soften. Rainforest and coral reefs fascinate us, the endless pine steppes of an industrial forest less so. Probably ecologists confronted with the task of assessing the diversity of an ecosystem could renounce complicated sampling methods and simply trust what they see, smell and hear. In the world of living beings the beautiful system most often is the diverse system, and the diverse system is the good system because life imagines itself as the greatest possible plenitude. Still, the beauty of natural systems never appears in the radiant triumph of victory. Ecological stability and the beauty of life are built on the dialectics of birth and death. It is fragile to the core. Its beauty, to which we are free to contribute at any moment, is the hope for healing.

Acknowledgments

I am enthusiastic that this book, my English language rewrite and update of the 2007 German volume "Alles fühlt," is now accessible to an English language audience. As most of my inspiration comes from the Anglo-American context of cognitive sciences, nature writing and ecophilosophy, I now can give back some fruits of the thoughts and experiences that have nourished me over the last decade. A huge thanks to the editors at NSP, particularly to Heather Nicholas, for offering me such a beautiful welcome.

This book would not have been possible without the relentless commitment of my friend, fellow philosopher, author and renowned commons activist, David Bollier. David managed to find a safe harbor for the manuscript in a North American editorial landscape where I had no clue. He also copyedited the larger part of the first half in a meticulous and yet empathetic manner. David helped to carve out my thoughts in that particular way of his, by which I really grasp what I had intended to say. His friendship truly makes my life more wonderful.

Many other colleagues kindly lent a hand. Celeste Ceguerra and Amy Cohen Varela clarified a couple of chapters with gracious prose and poetic insight. Katherine Peil provided quick and generous assistance in making my English sound a bit more native. Her own work on the crucial role of emotions is deeply inspirational. Stuart Kauffman's enthusiasm towards my "enlivenment" hypothesis helps me persevere in an often difficult debate. Heartfelt thanks to David Abram for writing the Foreword and campaigning for the title that fits.

Heike Löschmann of the Heinrich Böll Foundation in Berlin invented the term "enlivenment" in an informal meeting in November 2012 and with this kicked off a new, integrated approach towards the enigma of life. Thanks also to Kalevi Kull who for years has been sympathetic to and supportive of my work. In his unique fashion of being straightforward and deep at the same time he coined the motto that says it all: "Love is everywhere with life. But to notice it makes more love, more aliveness."

Endnotes

FOREWORD

1. David Abram, *The Spell of the Sensuous: Perception and Language in a More-than-Human World*. New York, Pantheon Books, 1996; p. 50.

2. Weber is careful not to draw too hard and fast a line between living beings and an ostensibly inanimate matter which forms the passive backdrop against which life unfolds. For he is attentive to the growing body of evidence that matter is self-organizing from the get-go, and hence that life, and its concurrent subjectivity, may be a tendency inherent in bare existence per se.

INTRODUCTION

1. Edward O. Wilson, *The Social Conquest of Earth*, Boston: W.W. Norton & Company, 2012.

2. Andreas Weber, *Enlivenment: Towards a Fundamental Shift in the Concepts of Nature, Culture, and Politics*, Berlin: Heinrich-Böll-Stiftung, 2013. Download at: http://www.boell.de/sites/default/files/enlivenment_v01.pdf

3. David Rudrauf, personal communication, 24 May 2006.

4. Edward O. Wilson and Steven R. Kellert, *The Biophilia Hypothesis*, Washington D.C.: Island Press, 1995.

5. Henry Miller, *Wisdom of the Heart*, New York: New Directions, 1960, 24.

6. Andreas Weber, "Was ist so schön an der Natur?" *GEO*, 1997.

7. Richard Louv, *Last Child in the Woods: Saving Our Children from Nature-Deficit Disorder*, Chapel Hill: Alonquin Books, 2005.

8. Miller, *Wisdom*.

9. The concepts of "poetic objectivity" and "empirical subjectivity" are central to the Enlivenment hypothesis. I explore them in depth in Weber, *Enlivenment*.

CHAPTER 1

1. Stewart Kauffman, *At Home in the Universe: The Search for Laws of Complexity*, London: Penguin, 1995.

2. Gerald von Dassow, E. Meir, E.H. Munro and G.M. Odell, "The segment polarity network is a robust developmental module," *Nature 406* (2000), 188–192.

3. Marc Kirschner, J. Gerhart and T. Mitchison, "Molecular 'Vitalism,'" *Cell* 100, (2000), 79–88.

4. Rick A. Relyea, "The Lethal Impacts of Roundup and Predatory Stress on Six Species of North American Tadpoles," *Archives of Environmental Contamination and Toxicology*, Vol. 48. No. 3, (2005), 351–357. "New effects of Roundup on amphibians: Predators reduce herbicide mortality; herbicides induce antipredator morphology," Ecological Applications, Vol. 22 Issue 2, (2012), 634–647.

CHAPTER 2

1. William Blake, *Auguries of Innocence and Other Lyric Poems*, Create Space Independent Publishing Platform, 2014, 7.

2. A list of biologists working on the idea of a new understanding of evolutionary theory can be found online at thethirdwayofevolution.com/people.

3. Francisco J. Varela, "Patterns of Life: Intertwining Identity and Cognition," *Brain and Cognition*, 37, (1997), 72–84.

4. Von Dassow et al., "Segment polarity network."

5. Varela, "Patterns of Life."

CHAPTER 3

1. Kauffman, *At Home*.

2. Bruce Alberts, quoted in Elizabeth Fox Keller, *Making Sense of Life: Explaining Biological Development with Models,*

Metaphors and Machines, Cambridge: Harvard University Press, 2002, 230.

3. Sean Carroll, *Endless Forms Most Beautiful: The New Science of Evo Devo*, New York: W. W. Norton, 2005.

4. Marc W. Kirschner and John C. Gerhart, *The Plausibility of Life*, New Haven and London: Yale University Press, 2005, 111.

5. Quoted in Fox Keller, *Making Sense*.

6. Robert D. Denham, "Interpenetration as a Key Concept in Frye's Critical Vision," in *Rereading Frye: The Published and the Unpublished Works*, David V. Boye and Imre Salusinszky, eds., Toronto: University of Toronto Press, 1999, 141.

CHAPTER 4

1. Anton Markoš et al., *Life as its Own Designer: Darwin's Origin and Western Thought*, 2009, 85.

2. Quoted in Stefan Schmitt, "Grünzeug mit Grips," *Die Zeit* 4, (2005).

CHAPTER 5

1. David Rudrauf and Antonio Damasio, "A Conjecture Regarding the Biological Mechanism of Subjectivity and Feeling," *Journal of Consciousness Studies* 12, no. 8–10, (2005), 237.

2. Jaak Panksepp, "Emotions as Viewed by Psychoanalysis and Neuroscience: An Exercise in Consilience," *Neuropsychoanalysis: An Interdisciplinary Journal for Psychoanalysis and the Neurosciences*, 1, (1999) 15–38.

3. Panksepp, *ibid.*

4. Michel Le Van Quyen, personal communication.

5. Varela, *Patterns of Life*.

6. Rudrauf and Damasio, "Conjecture."

7. Gaston Bachelard, *Poetics of Space*, Boston: Beacon, 1994.

8. George Lakoff and Mark Johnson, *Metaphors We Live By*, Chicago: Chicago University Press, 1980, 85.

9. George Lakoff and Mark Johnson, *Philosophy in the Flesh*, New York: Basic Books, 1999.

CHAPTER 6

1. This argument follows Howard Morphy, *Ancestral Connections: Art and an Aboriginal System of Knowledge*, Chicago and London: University of Chicago Press, 1991.

2. Wilson and Kellert, *Biophilia*; Paul Shepard, *Thinking Animals: Animals and the Development of Human Intelligence*, Athens, Georgia: University of Georgia Press, 1998.

3. The French anthropologist Claude Lévi-Strauss used this term to describe our way of making sense of the world, which is deeply embedded in our biological nature. Claude Lévi-Strauss, *The Savage Mind*.

CHAPTER 7

1. Andreas Weber, *Lebendigkeit: Eine Erotische Ökologie*, München: Kösel, 2014.

2. Andrew N. Meltzoff and M. Keith Moore, "Imitation of Facial and Manual Gestures by Human Neonates," *Science*, vol. 198 no. 4312 (1977), 75–78.

3. Jerry A. Fodor, *Psychosemantics: The Problem of Meaning in the Philosophy of Mind*, Cambridge, MA: MIT Press, 1987, 132.

4. Julia Simner, et al., "Synaesthesia: The Prevalence of Atypical Cross-Modal Experiences," *Journal of Perception*, 35, (2006), 1024–1033.

5. Andrew N. Meltzoff, "Imitation and Other Minds: The 'Like-Me-Hypothesis,'" in *Perspectives on Imitation: From Neuroscience to Social Science*, vol. 2, S. Hurley and N. Chater, eds., Cambridge, MA: Harvard University Press, 2005, 55–77.

6. Meltzoff, *ibid.*

7. Evan Thompson, "Empathy and Consciousness," in *Between Ourselves: Second-Person Issues in the Study of Consciousness*, ed., Evan Thompson, ed., Charlottesville, VA: Imprint Academic, 2001, 4.

8. Giacomo Rizzolatti, Luciano Fadiga, Vittorio Gallese, Leonardo Fogassi,

"Premotor cortex and the recognition of motor actions," *Cognitive Brain Research*, 3 (1996) 131–141.

9. Paul Valéry, *Tel Quel*, Paris: Gallimard, 1941, 42. My translation.

10. Meltzoff, "Imitation."

11. John Bryant, "To fight some other world," in *Ungraspable Phantom: Essays on Moby Dick*, John Bryant, Mary Bercaw Edwards and Timothy Marr, eds., Kent OH: Kent State University Press, 2006.

CHAPTER 8

1. Samuel Taylor Coleridge, *Selected Poetry*, London: Penguin Books, 1996, 57.

2. John Clare, in David Rothenberg, *Why Birds Sing: A Journey into the Mystery of Birdsong*, New York: Basic Books, 2005, 25.

3. Amotz Zahavi, "Mate selection—A selection for a handicap," *Journal of Theoretical Biology*, 53, Issue 1 (1975), 205–214.

4. Suzanne K. Langer, *Philosophy in a New Key: A Study in the Symbolism of Reason, Rite, and Art*, Cambridge, MA.: Harvard University Press, 1942.

5. Ivy Campbell-Fisher, "Aesthetics and the Logic of Sense," *Journal of General Psychology*, 43, Issue 2 (1950), 245–273.

CHAPTER 9

1. Algernon Charles Swinburne, *Songs of the Springtides and Birthday Ode*, London: William Heinemann, 1917.

2. Weber, *Enlivenment: Towards a Fundamental Shift in the Concepts of Nature, Culture, and Politics*, Berlin: Heinrich-Böll-Stiftung, 2013.

3. John S. Rosenberg, "Of Ants and Earth," *Harvard Magazine*, March-April 2003, 36–41.

4. Stephen J. Gould and Richard Lewontin, "The Spandrels of San Marco and the Panglossian Paradigm: A Critique of the Adaptionist Programme," *Proceedings of the Royal Society of London*, Series B, 205, No. 1161, (1979) 581–598.

5. Filip Jaroš, "The Ecological and Ethological Significance of Felid Coat Patterns (Felidae)." Doctoral thesis submitted to Charles University, Prague, Department of Philosophy and History of Science, 2012.

6. Cormac McCarthy, *The Border Trilogy*, Everyman's Library, 1999.

7. Geerat J. Vermeij, *Nature: An Economic History*, Princeton: Princeton University Press, 2004, 314.

8. Jon Young, personal communication, 21 February 2015.

9. Jules Michelet, *La Mer*, in *Oeuvres complètes*, Vol. 29, Paris: Hachette, 2013 (1893).

10. Adolf Portmann, *Animal Form and Patterns: A Study of the Appearance of Animals*, New York: Schocken Boooks, 1967.

11. Albert Camus, *Notebooks 1935–1942*, Chicago: Ivan Dee, 2010.

12. Kauffman, *At Home*.

13. Humberto Maturana and Francisco Varela, *The Tree of Knowledge: The Biological Roots of Human Understanding*, Boston: Shambhala, 1987.

14. Bruce H. Weber and David J. Depew, "Natural Selection and Self-Organization: Dynamical Models as Clues to a New Evolutionary Synthesis," *Biology and Philosophy*, 1 (1996) 33–65.

15. Daniel R. Brooks and Edward O. Wiley, *Evolution as Entropy: Towards a Unified Theory of Biology*, Chicago: University of Chicago Press, 1986.

16. Stuart Kauffman, *Investigations*, Oxford: Oxford University Press, 2000.

17. Kauffman *ibid*.

18. Michael L. Rosenzweig, *Species Diversity in Space and Time*, Cambridge: Harvard University Press, 1995.

19. Bruno Latour, *We Have Never Been Modern*, Cambridge: Harvard University Press, 1993, 85.

20. Andreas Weber, "Plankton," *Greenpeace Magazine*, 3, 2005.

CHAPTER 10

1. Albert Camus, *Summer*, London: Penguin, 1995.

2. William Beebe, *The Bird: Its Form and Function*, New York: Henry Holt, 1906.

CHAPTER 11:

1. Farooq Azam, in Robert Kunzig, *Mapping the Deep: The Extraordinary Story of Ocean Science*, New York: W. W. Norton, 2000, 207.

2. Konstantin Mereschkowsky, "The Theory of Two Plasms as the Basis of Symbiogenesis, a New Study or the Origins of Organisms," *Biol Centralbl.* 30, (1910) 353–367. Lynn Sagan, "On the origin of mitosing cells," *Journal of Theoretical Biology*, 14, March 1967, 255–274.

3. Lynn Margulis and Dorion Sagan, *What is Life?* New York: Simon & Schuster, 1995.

4. Peter J. Turnbaugh, et al., "The Human Microbiome Project," *Nature* 449, (2007), 804–810.

5. Lynn Margulis and Dorion Sagan, *Acquiring Genomes: A Theory of the Origins of Species*, New York: Basic Books, 2002, 7.

6. Personal communication. See also Katherine T. Peil, "Emotion: The Self-Regulatory Sense," *Global Advances in Health and Medicine*, 3 (2), 2014, 80–108.

7. Victor von Weizsäcker, in Annelie Keil, *Auf brüchigem Boden Land gewinnen: Biografische Antworten auf Krankheit und Krisen*, Munich: Kösel, 2011. My translation.

CHAPTER 12

1. Matthew Arnold and Nicole D. Fogarty, "Reticulate Evolution and Marine Organisms: The Final Frontier?" *International Journal of Molecular Science*, 10 (2009), 3836–3860.

2. Donald I. Williamson, "Caterpillars evolved from onychophorans by hybridogenesis," *PNAS* Vol. 106 no. 47, (2009) 19901–19905.

3. M. M. Martis et al., "Reticulate Evolution of the Rye Genome," *The Plant Cell*, 25 (2013), 3685–3698.

4. Donald I. Williamson, in Margulis and Sagan, *Acquiring Genomes*, 172.

5. *Ibid.*

6. *Ibid.*, 165.

7. For an overview of the controversy see Anna Marie A. Aguinaldo et al., "Evidence for a clade of nematods, arthropods, and other moulting animals," *Nature* 387 (1997) 489–439; Jennifer Grenier et al., "Evolution of the entire arthropod Hox gene set predated the origin and radiation of the onychophoran/arthropod clade," *Current Biology*, vol. 7 no. 8, (1997), 547–553; Donald I. Williamson, "Caterpillars evolved from onychophorans by hybridogenesis," *PNAS*, vol. 106 no. 47 (2009), 19901–19905; Albert D. G. de Roos, "Origin of insect metamorphosis based on design-by-contract: larval stages as an atavism," (2006) Online at: albertderoos.nl/publications/molecular-evolution; Jerome C. Regier, "Arthropod relationships revealed by phylogenomic analysis of nuclear protein-coding sequences," *Nature*, 463, (2010), 1079–1083.

8. François Jacob, in Carroll, *Endless Forms*.

9. Margulis and Sagan, *Acquiring Genomes*, 194.

10. Eva Jablonka and Marion J. Lamb, *Evolution in Four Dimensions: Genetic, Epigenetic, Behavioral, and Symbolic Variation in the History of Life*, Cambridge, MA: MIT Press, 2005.

11. Katherine Peil, personal communication.

CHAPTER 13

1. Weber, *Enlivenment: Towards a Fundamental Shift in the Concepts of Nature, Culture, and Politics*, Berlin: Heinrich-Böll-Stiftung, 2013.

2. J. M. Coetzee, *The Lives of Animals*, Princeton: Princeton University Press, 1999, 53–54.

3. Robert Costanza et al., "The value of the world's ecosystem services and natural capital," *Nature* 387, (1997), 253–260.

4. Andreas Weber, "Biodiversität: Die Natur als Wirtschaftsfaktor," *GEO* 5, 2008.

5. Weber, *Enlivenment*.

6. Gregory Bateson, *Steps to an Ecology of the Mind: A Revolutionary Approach to Man's Understanding of Himself*, Chicago: University of Chicago Press, 1972, 271–278.

7. Frans de Waal, *Primates and Philosophers. How Morality Evolved*, Princeton: Princeton University Press, 2006, 56.

CHAPTER 14

1. Johann Wolfgang von Goethe, in Rudolf Steiner, *Goethe's Conception of the World*, London: Kessinger, 2003.

2. Gilles Deleuze and Félix Guattari, *A Thousand Plateaus: Capitalism and Schizophrenia*, Minneapolis: University of Minnesota Press, 1987, 5.

3. From Leonard Cohen's song "Anthem": "There is a crack in everything / That's where the light comes in."

4. Andreas Weber, "The Economy of Wastefulness: The Biology of the Commons," in *The Wealth of the Commons: A World Beyond Market and State*, David Bollier and Silke Helfrich, eds., Amherst: Levellers Press, 2012, 6–11.

5. David Bollier, *Think Like a Commoner: A Short Introduction to the Life of the Commons*, Gabriola Island, BC: New Society Publishers, 2013, 174.

6. Edgar Morin, *L'identité humaine. La methode 5. L'humanité de l'humanité*, Paris: Seuil, 2001, 272.

7. Neera M. Singh, "The affective labor of growing forests and the becoming of environmental subjects: Rethinking environmentality in Odisha, India," *Geoforum*, 47, (2013), 189–198.

8. Andreas Weber, "Reality as Commons: A Poetics of Participation for the Anthropocene," in *Patterns of Commoning*, David Bollier and Silke Helfrich, eds., Amherst: Levellers Press, 2015.

9. Weber, *Enlivenment*.

10. Sandra Lubarsky, "Living Beauty," in *Keeping the Wild: Against the Domestication of the Earth*, George Wuerthner, Eileen Crist, Tom Butler, eds., Washington, DC: Island Press, 2014, 194–195.

Index

About the Author

Dr. Andreas Weber is a writer and independent scholar based in Berlin and Italy. He has degrees in Marine Biology and Philosophy, having collaborated with theoretical biologist Francisco Varela in Paris. Andreas' work is focusing on a re-evaluation of our understanding of the living. He is proposing to understand organsims as subjects and hence the biosphere as a meaning-creating and poetic reality. His long standing interest is how human feeling, subjectivity and aesthetics are related to biological realities. He proposes a Poetics of Nature to overcome the mind-body gap and to stop treating the biosphere as dead matter. He has put forth his ideas in several books translated into major languages and has contributed to newspapapers and magazines, such as *GEO* and *Die Zeit*. Weber regularly lectures and teaches at various workshops. He teaches at Leuphana University Lüneburg and at the University of the Arts, Berlin.

If you have enjoyed *The Biology of Wonder* you might also enjoy other

BOOKS TO BUILD A NEW SOCIETY

Our books provide positive solutions for people who want to make a difference. We specialize in:

**Food & Gardening • Resilience • Sustainable Building
Climate Change • Energy • Health & Wellness • Sustainable Living
Environment & Economy • Progressive Leadership • Community
Educational & Parenting Resources**

New Society Publishers

ENVIRONMENTAL BENEFITS STATEMENT

New Society Publishers has chosen to produce this book on recycled paper made with **100% post consumer waste,** processed chlorine free, and old growth free.

For every 5,000 books printed, New Society saves the following resources:[1]

38	Trees
3,436	Pounds of Solid Waste
3,781	Gallons of Water
4,931	Kilowatt Hours of Electricity
6,246	Pounds of Greenhouse Gases
27	Pounds of HAPs, VOCs, and AOX Combined
9	Cubic Yards of Landfill Space

[1]Environmental benefits are calculated based on research done by the Environmental Defense Fund and other members of the Paper Task Force who study the environmental impacts of the paper industry.

For a full list of NSP's titles, please call 1-800-567-6772 *or check out our website* at:

www.newsociety.com